Balancing Agile Combat Support Manpower to Better Meet the Future Security Environment

Patrick Mills, John G. Drew, John A. Ausink, Daniel M. Romano, Rachel Costello

RAND Project AIR FORCE

Prepared for the United States Air Force
Approved for public release; distribution unlimited

The research described in this report was sponsored by the United States Air Force under Contract FA7014-06-C-0001. Further information may be obtained from the Strategic Planning Division, Directorate of Plans, Hq USAF.

Library of Congress Control Number: 2014940914

ISBN: 978-0-8330-8208-4

The RAND Corporation is a nonprofit institution that helps improve policy and decisionmaking through research and analysis. RAND's publications do not necessarily reflect the opinions of its research clients and sponsors.

Support RAND—make a tax-deductible charitable contribution at www.rand.org/giving/contribute.html

RAND® is a registered trademark.

Preface

During Operation Iraqi Freedom and Operation Enduring Freedom, the U.S. Air Force (USAF) experienced demands for agile combat support (ACS) that were unique in size, composition, and duration, putting stress on a number of career fields. Out of concern for those and future operations, the Air Force Director of Transformation, Deputy Chief of Staff for Logistics, Installations, and Mission Support (AF/A4I) asked RAND Project AIR FORCE to analyze how well the USAF's ACS forces are postured to meet emerging requirements and how they might be better postured to meet those requirements. That request initiated a multiyear body of research into the ability of ACS forces to meet the needs of both traditional and emerging missions, such as joint expeditionary taskings (JETs), joint overseas base support, humanitarian assistance and disaster relief, counterinsurgency (COIN) operations, and building partnerships.

This report continues and extends that analysis by assessing the USAF's total force ACS manpower mix and identifying how to better prepare the USAF for future scenarios. Our analysis undertook four main objectives: assess the supply of ACS forces; assess future expeditionary demands; assess home-station requirements; and assess policy options for better shaping the mix of ACS forces to meet future demands, while accounting for their ongoing home-station mission.

This report should be of interest to operational planners, manpower planners, and analysts and planners in the ACS community and in particular could inform current force-shaping efforts.

This research was commissioned by AF/A4I and the Commander, Air Force Materiel Command, representing the command's role as the ACS Core Function Lead Integrator (CFLI). The analysis was conducted within the Resource Management Program of RAND Project AIR FORCE.

RAND Project AIR FORCE

RAND Project AIR FORCE (PAF), a division of the RAND Corporation, is the U.S. Air Force's federally funded research and development center for studies and analyses. PAF provides the Air Force with independent analyses of policy alternatives affecting the development, employment, combat readiness, and support of current and future air, space, and cyber forces. Research is conducted in four programs: Force Modernization and Employment; Manpower, Personnel, and Training; Resource Management; and Strategy and Doctrine.

Additional information about PAF is available on our website:
http://www.rand.org/paf/

iii

Contents

Figures

Table

Summary

The U.S. Air Force's (USAF's) current approach to sizing and shaping non-maintenance agile combat support (ACS) manpower often results in a discrepancy between the supply of ACS forces and operational demands. While requirements for operators and maintainers are generally driven by operational plans, numbers of platforms, and workload models, non-maintenance ACS career fields are often driven solely by home-station requirements (e.g., security forces [SF]). The net result is that much of ACS is sized and shaped to meet the requirements of home-station installation operations, not expeditionary operations.

We propose a more enterprise-oriented approach to determining ACS manpower requirements that addresses the imbalances that result from the USAF's current approach. Our approach sets total force ACS manpower (active, reserve, and civilian) at levels that can meet a range of expeditionary demands while maintaining home-station support.

Our approach maximizes expeditionary ACS capacity in terms of the ability to support two types of operations: **steady-state** (i.e., rotational) operations, in terms of the number of steady-state bases provided by the active-duty force; and **surge** (i.e., major combat) operations, in terms of the number of surge bases provided by the total force. We developed these base-driven metrics of expeditionary capacity by synthesizing combatant commander (COCOM) operational plans (OPLANs), Defense Planning Scenarios, functional area deployment rules, and subject-matter expert (SME) input. This maximization of expeditionary capacity is subject to the constraint of keeping home-station manning levels (active duty and civilians) constant.

We used these new expeditionary metrics to assess the capacity of the current ACS manpower mix to support expeditionary operations and found that there are imbalances among its career fields relative to expeditionary demands (i.e., the relative proportions of career-field capacity and skill mix do not match those found in planning scenarios). To remedy these imbalances, we developed and assessed several rebalanced manpower mixes to increase expeditionary capacity for both steady-state and surge operations while maintaining home-station support.

We found the current capacity of the limiting resources within ACS to be seven steady-state bases (with an active-duty deploy-to-dwell ratio of 1:2) and 22 surge bases (with the total force). Our first alternative ACS manpower mix rebalanced expeditionary capacity within *current end strengths*. This manpower mix has a steady-state capacity of 21 bases and a total-force surge capacity of 65 bases, with a small savings of $1 million per year. Alternative 2 takes a 25 percent cut in expeditionary capacity, with 16 steady-state bases and 49 surge bases, and offers savings of $340 million per year. Alternative 3 takes a 50 percent cut in capacity, with 11 steady-state bases and 33 surge bases, and offers savings of $670 million per year. All three alternatives have more expeditionary capacity than the limiting resources in the current ACS manpower mix.

In sum, the USAF can achieve more expeditionary ACS capacity than it currently has by realigning manpower, and it can realize substantial savings by reducing end strength and substituting civilian billets for military billets. To realize these benefits in a sustainable way, the entire ACS enterprise needs to strategically align forces to meet both sets of demands based on a common set of requirements and assumptions. We see two key obstacles to that transition.

First, ACS authority is decentralized and unintegrated, hindering a concerted effort to balance ACS functional capabilities with an enterprise view. Functional communities hold local expertise about their career fields and ultimately some level of authority, as they set policy. Major command (MAJCOM) commanders are charged with the organize, train, and equip responsibilities that contribute to ACS capabilities over the long term. For programming, the ACS Core Function Lead Integrator (CFLI) has the task of integrating inputs across ACS, while programming panels actually represent ACS within the corporate structure. As a result, the Vice Chief of Staff of the Air Force is the lowest-ranking person with the purview to arbitrate competing demands to balance capabilities across all ACS functional capabilities, simply by virtue of his position.

Second, there is inadequate operational and ACS policy to inform the manpower system as to how to shape ACS forces. There has not been an official USAF strategic plan released in over five years. The Annual Planning and Programming Guidance (APPG), which has historically included some relevant instruction, has varied in its level of detail regarding specific planning scenarios and the degree of guidance it has given to programmers about how to trade off competing objectives when resources cannot support the full range of stated planning objectives. The Air Force Strategic Master Plan (a product established at the fall 2013 CORONA) and the ACS CFLI's Core Function Support Plan could set guidance around which to shape ACS forces, but because the CFLI currently lacks official directive authority over the ACS enterprise, any analysis the ACS CFLI presents will require broad consensus among the above stakeholders to be implemented.

What is needed is a clear articulation of *operational* objectives and priorities that can be translated into quantifiable *ACS* objectives that will drive the size and shape of the future force. To that end, we recommend several steps for implementing the concepts laid out in this report.

First, the USAF should provide clearer strategic planning guidance about future objectives to direct planners and programmers in sizing and shaping the force. Second, those clearly stated objectives should be translated into measurable ACS objectives (e.g., the type and number of expeditionary bases to support).

Third, the ACS manpower system should broaden its planning objectives to include both expeditionary and home-station requirements. Fourth, the USAF should specify policies to inform manpower trade-offs when issues arise in balancing competing needs (e.g., expeditionary vs. home-station needs, an enterprise vs. a career-field perspective).

The changes we recommend are substantial and would require years to fully implement. However, until they are implemented, the ACS enterprise will continue to produce a set of capabilities that are neither balanced amongst themselves nor against operational needs.

Acknowledgments

This work could not have been completed without the support of Mr. Grover Dunn, former AF/A4I, and Maj Gen (ret) David Eidsaune, former AFMC/A8/9, the analysis sponsors. Col Steven Lawlor, Mr. Dunn's deputy, was an extremely helpful action officer who always provided excellent support for our research. Maj Alex DeVoe, Mr. Dunn's executive officer, was also of great help.

Many functional experts on the Air Staff and in the broader USAF community provided data, documents, interviews, and helpful feedback. They are too numerous to name here, but we thank them for opening their doors to us and providing much valuable support.

Within RAND, we thank many colleagues for their research contributions. We thank Manuel Carrillo and Gary Massey for providing data and patiently explaining the inner workings of USAF manpower and personnel systems. We thank Rena Rudavsky for her data analysis. For their discussions, suggestions, and feedback about this work, we thank Ron McGarvey, Al Robbert, Anthony Rosello, Don Snyder, and Bob Tripp. We thank Kristin Leuschener for helping to edit and structure the report. Regina Sandberg provided helpful editing and administrative support. We thank Don Snyder and Al Robbert for their helpful reviews of this report.

The authors take responsibility for any errors or omissions in this report.

Abbreviations

A2AD	anti-access/area denial
ACS	agile combat support
AD	active duty
AF/A4I	Air Force Director of Transformation, Deputy Chief of Staff for Logistics, Installations, and Mission Support
AFI	Air Force Instruction
AFMS	Air Force Manpower Standard
AFPC	Air Force Personnel Center
AFPD	Air Force Policy Directive
AFSC	Air Force Specialty Code
AGR	Air Guard Reserve
AOR	area of responsibility
APPG	Annual Planning and Programming Guidance
ARC	air reserve component
ART	Air Reserve Technician
BOS-I	base operating support—integration
BP	building partnerships
CBRN	chemical, biological, radiological, and nuclear
CENTCOM	Central Command
CFLI	core function lead integrator
CFSP	Core Function Support Plan
COCOM	combatant commander
COIN	counterinsurgency
CONOP	concept of operations
D2D	deploy-to-dwell
DHP	Defense Health Program
DoD	Department of Defense
EM	emergency management
EMEDS	expeditionary medical support
EOD	explosive ordnance disposal
FES	firefighting emergency services
FY	fiscal year

ISC	integrated security construct
ISP	Integrated Security Posture
JET	joint expeditionary tasking
LCOM	Logistics Composite Model
MAJCOM	major command
MCO	major combat operation
OPLAN	operational plan
OSD	Office of the Secretary of Defense
PAF	Project AIR FORCE
PFP	Planning Force Proposal
PRIME BEEF	Prime Base Engineer Emergency Force
QDR	Quadrennial Defense Review
RED HORSE	Rapid Engineer Deployable Heavy Operational Repair Squadron, Engineer
SF	security forces
TFE	Total Force Enterprise
UMD	unit manning document
USAF	United States Air Force
UTA	UTC Availability
UTC	Unit Type Code

1. Introduction

The U.S. Air Force (USAF) is at a potential turning point. It has been engaged in expeditionary operations for more than two decades, developing and adapting its force management concepts to cope with changing demands first in Iraq, then in Afghanistan, and then again in Iraq, as well as elsewhere. As those operations wind down, so do some of the demands that have strained the force for many years, potentially leaving a window within which to reset and prepare for future operations.

Recent defense guidance has shifted the focus toward Asia, emphasizing missions that may stress USAF capabilities in new ways. The USAF is currently reexamining its force presentation and management construct, raising questions about how the force will position and prepare itself for the future. Additionally, fiscal pressures make resource trade-offs even more weighty and difficult.

The USAF has an opportunity both to reassess the size and shape of its forces and the policies it uses to govern them in light of potential future demands. However, this task is difficult, given the scope of USAF capabilities, the range of its missions, and the inherent complexity of the current security environment. This report describes and illustrates a method for sizing and shaping the USAF's agile combat support (ACS) forces to meet a range of future operations.[1] We present a framework that can be used to develop a robust force that can support both expeditionary and home-station commitments, and that makes explicit important assumptions and planning inputs.

Challenges to Shaping the Future ACS Force

The USAF is an expeditionary force, meaning that it fights from forward operating locations that generally do not house permanently stationed forces. These forward operating locations range in size, austerity of conditions (i.e., level of infrastructure), and severity of threat. To support the range of military operations at these various forward operating locations, the USAF maintains a wide array of manpower and equipment to deploy and employ forces rapidly.[2]

The USAF was organized, sized, and shaped for a more-traditional Cold War model of fighting,[3] which involved preparing to fight predominantly from garrison bases throughout

---

[1] In this report, we use *ACS* in the broad, enterprise sense, not merely to refer to the portion of the force defined by the current service core functions.

[2] In this report, we use *manpower* to refer to the billets or "spaces" that the USAF plans and budgets for, while we use *personnel* in reference to a base population being supported, e.g., "base personnel."

[3] An example of this Cold War model is wing-sized deployments with support forces sized and shaped to support the entire wing in major combat operations.

Europe and northeast Asia. Although it developed new capabilities and concepts to support the expeditionary operations that emerged in the 1990s, the USAF has not yet embraced structural changes that could better posture it to fight more-recent expeditionary operations.[4]

The 2012 Defense Strategic Guidance lists as primary missions for the U.S. armed forces a range of mission types, many of which could challenge the USAF.[5] These mission sets, which include counterinsurgency (COIN), building partnerships (BP), and major combat operations (MCOs) with anti-access/area denial (A2AD) threats, not only differ from one another in their demands but are also distinct in size and shape from the traditional model of warfighting around which the USAF is structured.[6] These missions sometimes entail operating in increasingly austere environments and under extremely high threats, often with dispersed forces, and sometimes outside the traditional boundaries of air bases. The impact of such conditions is felt most by the USAF's support forces, which the USAF calls *agile combat support*.

In addition to these mission sets, the USAF faces the specter of sustained commitments in the Central Command (CENTCOM) area of responsibility (AOR). Even though the war in Iraq has been concluded[7] and forces in Afghanistan are drawing down,[8] the USAF may yet play a significant role in regional stabilization in and around those countries for years to come.[9]

An additional factor complicating the planning of the USAF's resource mix for future operations is the rate at which the Office of the Secretary of Defense (OSD) plans and guidance change. Presidential administrations change every four or eight years, and each one rightly reassesses the geopolitical environment and the appropriateness of U.S. national security and military strategies. Administrations also respond to emerging changes in the security environment, and OSD appropriately responds to these changes, as well as foreseeable changes in threats, by issuing new guidance and planning scenarios around which the military services can plan and program.

---

[4] Raymond E. Conley, Albert A. Robbert, Joseph G. Bolten, Manuel Carrillo, and Hugh G. Massey, *Maintaining the Balance Between Manpower, Skill Levels, and PERSTEMPO*, Santa Monica, Calif.: RAND Corporation, MG-492-AF, 2006. This report found that while the USAF was drawing down its forces in the 1990s and conducting more deployed operations, it did not change its manning to meet the simultaneous tasks of support deployments and sustaining garrison operations. It also found that during deployments, non-deploying personnel assigned to many functional areas (mostly within ACS) within the wings and commands were severely stressed and could not perform their normal home-base missions without working long hours. This was already true in the 1990s and early 2000s. Since then, the USAF has neither increased nor balanced ACS manpower to support these commitments.

[5] Barack Obama, *Sustaining U.S. Leadership: Priorities for the 21st Century*, Washington, D.C., January 2012.

[6] See United States Department of Defense, *Quadrennial Defense Review Report*, Washington, D.C., February 2010a; and United States Air Force, *USAF Strategic Planning 2010–2030 Strategic Environmental Assessment*, March 11, 2011b.

[7] Matt Negrin, "The Troops in Iraq: Sent Home, as Promised," ABCNews.com, July 7, 2012.

[8] "Obama Announces 34,000 Troops to Leave Afghanistan," *BBC News Online*, February 13, 2013.

[9] "Obama Announces 34,000 Troops to Leave Afghanistan," 2013.

OSD utilizes a cyclical process by which it creates, releases, and further refines a range of planning scenarios, including those for steady-state (i.e., rotational) operations, surge (i.e., major combat) operations, post-surge demands, small-scale contingencies, and irregular wars. New scenarios are released almost monthly, and even overarching force-shaping constructs change within a single administration.

As an example, in 2009, when the Obama administration took office, OSD's force-sizing construct was called the *Integrated Security Posture* (ISP). This construct contained sets of surge and steady-state scenarios arranged to create a potential future against which military planners could assess their force structure (and hence resize and shape them). In 2010, the Quadrennial Defense Review (QDR) included three different sets of scenarios (later called *integrated security constructs* [ISCs]) against which the services were directed to size and shape their forces.[10] Because the details of those scenarios did not yet exist, OSD and service planners set about developing them. In 2012, before the last of the three ISCs was completed, the Obama administration released the new U.S. Defense Strategic Guidance, which again changed the strategic and operational objectives against which the military services were to size and shape their forces. To be clear, the Obama administration has merely exercised its proper authority over the military by changing its strategic and operational objectives. However, the evolving planning priorities create a difficult problem for the service planners, who must sustain a healthy force while responding to the changing security environment.

A final issue affecting planning for ACS manpower is budget. In spite of the increasing demand for ACS forces, the USAF has reduced its ACS forces to underwrite the recapitalization of its aircraft fleets.[11] In addition, all the services face increasing budgetary pressures, including rising health care costs,[12] the Budget Control Act of 2011 caps on annual appropriations, and the automatic spending reductions that took effect at the beginning of March 2013.[13]

Even in less-lean times, it was a complex task for the USAF to balance its portfolio of ACS manpower and equipment to best support its array of operational missions, but this task proves even more challenging today.

Project AIR FORCE Research to Assess the USAF ACS Manpower Mix

RAND Project AIR FORCE (PAF) began conducting research to inform these issues several years ago. In fiscal year (FY) 2008, PAF conducted research to estimate emerging future

---

[10] U.S. Department of Defense, 2010a.

[11] Rodney McKinley, *Roll Call*, January 12–16, 2007.

[12] Robert M. Gates, "Opening Summary — House Appropriations Committee–Defense (Budget Request)," Washington, D.C., March 2, 2011; and Congressional Budget Office, *Long-Term Implications of the 2012 Future Years Defense Program*, Washington, D.C., June 2011.

[13] Public Law 112–25, Budget Control Act of 2011, Section 365, 125 Stat. 240, 2011.

demands for USAF deployment capabilities across a range of mostly steady-state operations,[14] assess the USAF's ability to meet the demands of that security environment, and evaluate the potential of a few strategic policy levers to address manpower capability imbalances.[15] This initial research focused on midterm scenarios in which forces in Iraq and Afghanistan surged and then drew down, and it also explored what types of requirements indirect operations (those in which U.S. forces play a supporting role to indigenous forces or in which ACS is in a supported, rather than supporting, role) might hold for ACS forces.[16] One key finding of that report was that the then-current ACS manpower mix would not have been able to support mid-range defense steady-state OSD scenarios without significant shifts in manpower and/or major policy changes.

In FY 2010, PAF expanded on this research with a project that assessed the capacity of the ACS posture at the time to meet a range of future scenarios (both major combat and steady-state scenarios) for four key ACS functions associated with expeditionary basing: civil engineering, communications, medical, and security forces (SF). That research found significant imbalances between the manpower postured by those career fields and the demands of a range of near- and long-term MCOs.

This report expands that expeditionary analysis to other ACS functions and identifies policy options that can best prepare the USAF's ACS forces for the emerging security environment. While many aspects of the USAF's ACS resources are worthy of analysis in this context, the project sponsor directed our research team to focus first on manpower, as it consumes such a significant portion of the USAF's budget. For the USAF, one of the challenges inherent in determining expeditionary manpower requirements is the impact on home-station operations when personnel deploy. We address this challenge in our methodology, explained later.

Purpose and Scope of This Report

This report examines how the USAF can become better prepared for future operations by assessing alternative ACS manpower realignments. We describe a more enterprise-focused approach to determining the ACS manpower requirements needed to address the range of planning scenarios for which the USAF must be prepared. In this research, we sought to answer three key research questions:

---

[14] *Steady-state* generally refers to military deployments short of major combat operations, which are of extended duration, and are supported by forces on temporary duty, which rotate forward in successive deployments. The word *surge* is meant to signify major combat operations.

[15] Patrick Mills, David A. Shlapak, Ricardo Sanchez, and Robert S. Tripp, *Combat Support Beyond Iraq: Implications of the Future Security Environment for the USAF,* Santa Monica, Calif.: RAND Corporation, MG-1034-AF, 2011, not available to the general public.

[16] Examples of indirect operations explored in that research are humanitarian assistance, reconstruction, and ACS training and advising.

4

- What will future operations require of the ACS manpower mix? We begin by investigating what tasks ACS will perform in the future security environment.
- How well is the current ACS manpower mix prepared to meet the requirements of future operations? We then assess the ability of the current force to meet those requirements.
- What can the USAF do to better meet the requirements of future operations? Finally, we investigate an approach to mitigating the shortfalls we identified.

To make our analytic problem tractable, we narrowed our scope to include only a subset of ACS career fields that are critical to expeditionary basing and have a significant population. We included the following functional capability areas in our scope: SF; services; explosive ordnance disposal (EOD); fire protection; emergency management (EM), which handles chemical, biological, radiological, and nuclear (CBRN) threats;[17] engineering operations (i.e., Prime Base Engineer Emergency Force, or PRIME BEEF); contracting; fuels support; logistics readiness; expeditionary medical support (EMEDS);[18] and transportation (excluding air transport). Figure 1.1 shows the proportion of manpower included from the active duty (AD) and air reserve component (ARC).

---

[17] In this analysis, we include only emergency management capabilities managed by civil engineering, and exclude those managed by medical services, like bioenvironmental engineering.

[18] The scope of unit type codes (UTCs) we include for EMEDS can be found in Mills, 2012. It mostly includes EMEDS basic UTCs and preventive aerospace medicine for a 3,000-person deployment package.

Figure 1.1 Postured Manpower Included in This Analysis

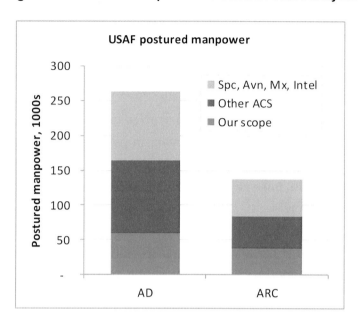

NOTE: Spc = space, Avn = aviation, MX = maintenance, Intel = intelligence.

The two columns in Figure 1.1 show the totality of AD and ARC manpower positions postured in the Unit Type Code Availability (UTA) data file.[19] The height of each column shows the postured manpower. Of the 263,000 AD manpower positions postured, about 60,000 positions are in our analytic scope (blue wedge). Another 99,000 are in operations and maintenance positions (green wedge), and thus would not be appropriate to include in our analytic approach (for reasons we will explain in the next chapter). Our analysis could be expanded to include many, but not all, of the remaining 104,000 or so positions (red wedge), most of which are in ACS. The proportions and applications are consistent for the ARC also.

Outline of This Report

The remainder of this report is divided into four chapters:

- In Chapter Two, we measure the expeditionary capacity of the current ACS force. We explain how we derived new metrics to measure that capacity, describe how we distilled current OSD guidance down to its key drivers of ACS demands, and show how well the current ACS force can support OSD plans.
- In Chapter Three, we explain how we developed alternative manpower mixes driven by expeditionary demands and show how well those alternative mixes perform against OSD plans.

---

[19] The entire USAF is not postured in the UTA. Institutional positions are excluded, and these make up the bulk of the 50,000-position gap between the numbers in the figure and the official USAF end strength.

- In Chapter Four, we discuss several additional factors that are not directly addressed in our analysis but that we feel are important for the USAF to consider.
- In Chapter Five, we offer some concluding thoughts, including a discussion of what would be necessary to implement the manpower realignments described in this report.

2. An Enterprise Approach to Determining ACS Manpower

In this chapter, we compare and contrast the USAF's current system for developing ACS manpower requirements with the more enterprise-focused approach we developed as part of our analysis. There are two main features of our approach that make it useful for exploring a range of alternative ACS manpower realignments: (1) it integrates both home-station and expeditionary requirements and (2) it provides an enterprise view of ACS.

Before describing this enterprise approach, we first describe the basic strategic planning context in which force-shaping decisions should be made and highlight the part of the system on which we focus. Second, we highlight several shortcomings in the USAF's current process for determining ACS manpower requirements.

Force-Shaping Decisions Must Consider Many Elements of the USAF

The USAF consists of many different elements, and long-term force-shaping actions must consider all of them to maintain an effective and efficient force. In Figure 2.1, we depict a simplified version of a strategic planning system to illustrate the various force elements and how one might approach shaping them. It is not meant to capture all aspects of the Air Force's planning system, but focuses on elements key to this discussion.

The top level of Figure 2.1 represents the setting of strategic and operational objectives, essentially the "ends" in the traditional ends-ways-means construct. From this set of objectives come the requirements for the rest of the force.

The center set of boxes represent expeditionary demands, both wing-level and regional or joint demands. In a contingency, much of the called-for USAF forces deploy to forward-operating locations within organizations that are wing-level and below. This usually entails aircraft, their operators, and support forces. That set of wing-level deployed forces creates a demand for regional support, whether staffs or theater-level assets such as Rapid Engineer Deployable Heavy Operational Repair Squadron Engineers (RED HORSE) or aeromedical evacuation.

Figure 2.1 Strategic Planning Context for Force-Shaping Decisions

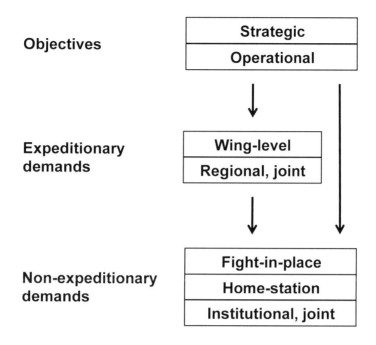

The bottom set of boxes in Figure 2.1 depicts non-expeditionary demands. It is sometimes the case that strategic and operational objectives directly drive demands for fight-in-place forces, such as space or missile squadrons. In other cases, those fight-in-place forces support, or work in concert with, particular forward-deployed forces (e.g., those permanently stationed on the Korean peninsula).

The remainder of the non-expeditionary forces, then, ought to be sized and shaped to support the range of demands above it (as pictured here). Home-station forces sustain garrison bases as necessary, and still more personnel support broader institutional (e.g., acquisition, research, and development) and joint requirements. All of these processes occur in the context of a range of legal, policy, and bureaucratic constraints.

In this report, we focus mainly on operational objectives and the expeditionary wing-level demands they drive. We treat the other components of the force (the boxes below wing-level expeditionary demands in Figure 2.1) as a kind of fixed cost and take them as a given in their current size and composition. When we speak of expeditionary forces, we exclude regional or joint demands from any calculation of supply or demand. Generally, a change in one element would create cascading changes to the elements below it, and a comprehensive assessment of the force would consider each of these in turn.

Given that strategic context, we now turn our attention to the USAF's current system for developing manpower requirements.

How the USAF Develops Manpower Requirements

The USAF uses a complex process to determine the personnel needed to perform its mission to fly, fight, and win in air space and cyberspace. At the highest level, Congress controls manpower by authorizing military "end strengths" (the number of personnel allowed in a military service), appropriating civilian work years, and "establishing military grade distributions and other guidelines."[20] The USAF manages its "human capital" through three intertwined subsystems:

- the manpower subsystem, which defines the demand for personnel and how it should be rationed among different components of the USAF
- the personnel subsystem, which manages the supply of personnel by placing people in appropriate positions
- the training subsystem, which trains and develops individuals to fill USAF jobs.[21]

Current USAF Process Uses Different Requirements and Approaches for Different Functional Communities

To determine the specific number of personnel needed for an organization to fulfill its given tasks, the USAF uses different methods for different functional communities. The number of operators is determined by the crew ratios required for aircraft, missile, and space systems. Thus, for example, the number of aircraft at a base will largely determine the number of pilots assigned there.[22]

The number of maintenance and munitions maintenance personnel is likewise driven by the type and number of platforms, but workload models are also factored in to ensure adequate system availability for the operators. Some maintenance manpower requirements are derived from simulations performed using the Logistics Composite Model (LCOM) or from aircraft-specific maintenance man-hour per flying hour requirements.[23] When using LCOM, the anticipated flying-hour program is an input, and the model uses actual component failure rates, repair times, and crew-size demands to estimate the manpower needed to satisfy the flying-hour program. This means that, to the degree that the number of platforms was originally derived from some set of strategic planning objectives, the operators and maintainers who support those platforms are tightly coupled with those objectives. This essentially follows the order laid out in Figure 2.1.

---

[20] Air Force Policy Directive (AFPD) 38-2, *Manpower and Organization: Manpower*, Washington, D.C., March 2, 1995.

[21] Conley et al., 2006.

[22] Crew ratios for Air Force aircraft are found in Air Force Instruction (AFI) 65-503, *U.S. Air Force Cost and Planning Factors*, Washington, D.C., February 4, 1994, Table A36-1, "Authorized Aircrew Composition."

[23] AFI 38-201, *Manpower and Organization: Management of Manpower Requirements and Authorizations*, Washington, D.C., September 26, 2011, pp. 13–19.

Non-maintenance ACS career fields determine their manpower differently. Most other ACS career fields utilize the Air Force Manpower Standard (AFMS), a "product/service-oriented" document that is meant to "quantify manpower resources required and the anticipated frequency or workload count for each product/service."[24] In practice, at a given USAF installation, values for the *workload drivers* are plugged into an equation to get the number of man-hours needed and *variances* that might be used to adjust the result.[25] For most ACS career fields, these workload drivers are based on home-station installation requirements, not expeditionary scenarios or other strategic objectives. This essentially reverses the strategic planning process pictured in Figure 2.1.

ACS career fields that do plan manpower in a way that attempts to integrate home-station and expeditionary requirements (e.g., medical services and civil engineering) do not use the same operational planning scenarios or requirements.[26]

There are two repercussions from the lack of alignment and integration among ACS career fields regarding manpower planning.

ACS Manpower Planning Typically Has a Career-Field Focus Rather Than an Enterprise Focus

First, much of ACS has a career-field focus rather than an enterprise focus. The individual functional managers do not look across functions or career fields to balance their capabilities with others, and there is no institutional mechanism in place to look across these career fields and the capabilities they represent to balance and shape the force as a whole.

Much of ACS Is Sized and Shaped to Meet Home-Station Requirements Rather Than Expeditionary Requirements

The second repercussion is that much of ACS is sized and shaped to meet the requirements of home-station installation operations, not expeditionary operations. One logical outcome of this approach is that the supply of ACS capabilities is inherently mismatched with the USAF's operational forces during deployments, and thus mismatched with operational needs. The USAF's operational forces (and the maintenance units that support them) are driven by operational objectives, and they shift over time in response to changing technologies, objectives, and threats. At the same time, these ACS career fields continue to be driven by the same home-

---

[24] AFI 38-201, 2011, para. 2.1.2.

[25] For example, Yokota Air Base in Japan has a variance that adds almost 2,000 man-hours to the calculation for contracting personnel. The total man-hour number is divided by "appropriate Man-hour Availability Factors (MAFs)" to determine the number of personnel authorized (per AFI 38-201). There might also be some "fixed" variances given in terms of the number of people that will increase the number of personnel. Again, using Yokota Air Base as an example, there is a variance of five additional personnel to take into account the added time to translate documents.

[26] The medical services use Joint Strategic Capabilities Planning Guidance, War Mobilization Planning Guidance, and Defense Planning Guidance, while civil engineering assesses historical deployment data, Air and Space Expeditionary Task Force (AETF) Force Modules, and OSD operational availability studies.

station installation requirements, which slowly shrink over time as bases are closed and manpower cuts reduce ACS forces.

Our conclusion is not that there is a problem with the USAF's process for developing manpower standards. Rather, the shortcoming we see is with the objectives the manpower system uses to set requirements. The current system sizes some career fields to home-station demands rather than expeditionary demands. Therefore, while the current manpower planning system may be adequate to meet peacetime home-station requirements, we seek an approach that balances long-range expeditionary demands with home-station operations.

How We Develop Manpower Requirements

We propose an alternative approach to developing manpower requirements that seeks to address the shortcomings of the USAF's current approach. Our approach has two key features: (1) it integrates both expeditionary and home-station requirements and (2) it takes an enterprise view of ACS. Figure 2.2 depicts our approach graphically.

Figure 2.2 Analytic Approach to Shaping ACS Manpower

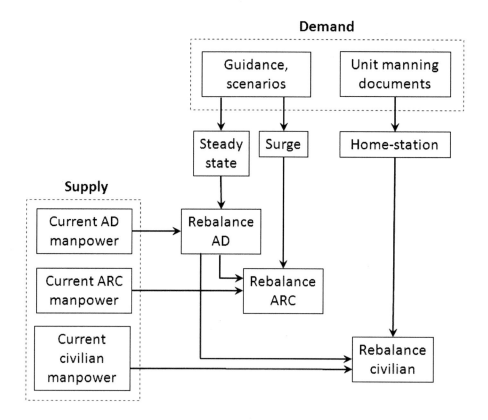

Supply of ACS Forces

The two key sets of inputs for our analysis are the supply of and demand for forces (i.e., manpower). For the supply of forces set presented on the left side of Figure 2.2, we obtained data on the current authorized manpower for AD, ARC, and civilians. We derived deployable military manpower from the UTA file, the primary inventory of the USAF's manpower capabilities. We took the postured manpower as it was stated in Unit Type Codes (UTCs) and Air Force Specialty Codes (AFSCs) and interpreted the coding of positions according to AFI 10-401.[27] If the UTA stated that forces were available for deployment, we assumed them to be so. If the UTA stated that they were not available, we assumed they were not. We acknowledge that the UTA has shortcomings. For example, some career fields have felt the need to increase UTC availability to meet day-to-day rotational demands, which overstates their true expeditionary capacity (relative to home-station impacts). But the UTA remains the best available data source of its kind.

We supplemented the UTA with military and civilian manpower data provided by the Air Staff to PAF.[28] For expeditionary operations, we decreased all career fields by an across-the-board planning factor of 20 percent, reasoning that some portion of the postured manpower would not be available for deployment, whether due to personnel assignment shortfalls, transfer, medical leave, or other reasons.[29]

Demand for ACS Forces

To quantify the demand for ACS forces, we departed from most traditional scenario-based manpower analyses. Most manpower analyses derive a demand signal from one or more operational scenarios, compare the supply of forces to this demand, and then quantify shortfalls in terms of unit types of personnel. We chose to diverge from this process for two reasons.

First, shaping ACS around one or a few specific scenarios could create a force wedded to the unique characteristics of those scenarios. How forces are bedded down has a significant impact on some ACS requirements, and basing opportunities can change over time as political alliances shift and threats change. Thus, we sought an approach that could capture the salient characteristics of future scenarios in the right time frame but that would not be wedded to any one scenario.

Second, we sought to develop a method of assessing and articulating ACS capabilities that would give ACS and operational planners a tool they could use to easily understand the operational implications of ACS resource limitations. We found that the results of more-

---

[27] AFI 10-401, *Air Force Operations Planning and Execution*, Washington, D.C., April 25, 2005.

[28] Air Force Personnel Center, Data Retrieval Section, Authorized Manpower Master file, September 30, 2010; Air Force Personnel Center, Data Retrieval Section, Active Enlisted End of Month Master Personnel Extract file, September 30, 2010.

[29] United States Air Force, *Annual Planning and Programming Guidance*, Washington, D.C., 2011a, not available to the general public.

traditional operational availability analyses that focus on ACS manpower or UTC shortfalls can be difficult to articulate in terms of operational objectives that could help guide an operator's decisionmaking.

To support these two aims, we developed metrics to express expeditionary ACS resource demands in operationally relevant terms. Figure 2.3 shows the inputs and process we used to develop these metrics and the outputs that result from each step.

Figure 2.3 Process for Developing Expeditionary ACS Metrics

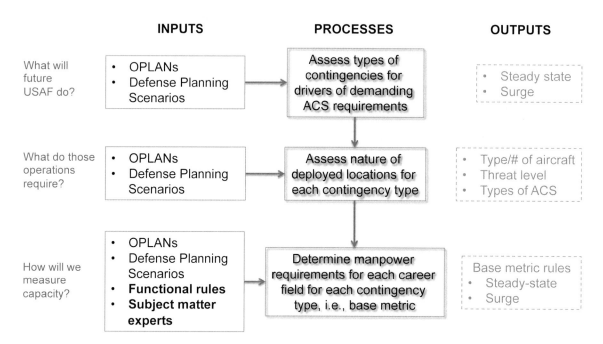

The first step, shown in the top row of Figure 2.3, began with a basic question of strategic planning: What will the future USAF do? To understand the sizing and shaping of the services, we reviewed guidance documents like the National Security Strategy, National Defense Strategy, National Military Strategy, the QDR, and other OSD guidance. We then looked to defense planning scenarios to better understand how these strategies translate into operational scenarios and specific requirements for military forces, especially the way in which the USAF intends to bed down its forces for different types of scenarios. We viewed these planning scenarios as a key articulation of future USAF demands.

We grouped these scenarios into broad categories based on similar attributes. To highlight the strategic choices that could lead the USAF leadership to reshape ACS, we focused our analysis on the types of contingencies that might put the most stress on the ACS force. The inputs to this process were OSD Defense Planning Scenarios and operational plan (OPLAN) time-phased force deployment data. We analyzed these scenarios to understand the types of contingencies the Department of Defense (DoD) and Service planners envision for the current year, as well as for the mid- to long term. The product of our assessment was to focus on two broad sets of contingency types: steady-state and surge operations. (For surge operations, we focused on long-range operations that would occur in the 2020s time frame.)[30]

The second step, shown in the middle row of Figure 2.3, was to assess the nature of the deployed locations for each contingency type. The motivation for this step was to see how engaging in one contingency type over another would change the requirement for support from ACS forces. The inputs for this step were the same scenarios as for the first step. The output was, for each scenario type, a rough characterization of the types and numbers of aircraft at deployed locations (which we distilled to the number of squadron equivalents per base), the threats to which they would be subject, and the types of ACS support that planners anticipated sending. The Appendix of this report explains in more detail the amount of variance there is in the wing-level beddowns for these scenario sets.

The final step, shown in the bottom row of Figure 2.3, was to determine the manpower requirements for each career field for each contingency type (i.e., a base metric). Because there is no authoritative source for expeditionary deployment rules, we gathered these data ourselves and synthesized an array of sources. The main inputs to this step were functional area deployment rules and subject-matter expert (SME) input (shown in bold in the bottom left-hand box of Figure 2.3). We synthesized these inputs to develop a standard set of ACS requirements for a base of each contingency type. To a lesser degree, we consulted the scenario data from the first two steps to provide a comparison with the deployment rules we developed. The details of these deployment rules are discussed in previous unpublished RAND research.

As the net result of this process, we measure the capacity of the ACS manpower mix to support two types of operations:

- **Steady-state operations.** This refers to *deployed* rotational operations, supporting expeditionary operating locations, and main operating bases that are operated during a contingency, often with additional deployed support. We measure steady-state capacity in terms of the *number of steady-state bases*. The basic steady-state base characteristics that drive ACS requirements are a total base population of 1,100 personnel, roughly one squadron of aircraft, a medium conventional threat level, and a low CBRN threat level.
- **Surge operations.** This refers to the maximum one-time deployment of all available military forces, including full mobilization of all reserve-component forces. These forces are deployed for the duration of the conflict and must be reset upon redeployment. We measure surge capacity in terms of the *number of surge bases*.[31] The basic surge-base characteristics that drive ACS requirements are a total base population of 1,650 personnel, roughly one and a half squadrons of aircraft, and high conventional and CBRN threat levels.

---

[30] We do not include joint expeditionary taskings in our force-shaping calculations, as these are not doctrinal USAF roles and missions for which the USAF is directed to plan and program.

[31] This usage of steady state and surge align with recent DoD guidance on sizing and shaping military forces.

We applied current functional deployment planning factors, as explained above, to derive manpower requirements for each career field according to the key drivers for each metric.

Because we shape (i.e., balance) ACS functional capabilities to the scenario sets we chose, ACS begins to look more like the demands found in those scenarios sets. Because we use metrics correlated to bases, decisionmakers can calibrate the amount of each capability desired to meet future operations, given the cost-capability trade-offs.

It is certainly possible to use a more sophisticated method than the one we describe above (e.g., to use different sizes of bases in a given contingency type, or to subdivide contingencies into more than just two types). However, we sought to keep this analysis as simple as possible to create a straightforward communication tool that enables the operational and support communities to dialogue about cost, capability, and risk at a high level, informed by rigorous analysis. We intend this analysis to complement other analyses that focus on specific scenarios or use more-sophisticated modeling.

Ultimately, these two base metrics have two key characteristics: size and shape. The size of each base metric is simply the total number of ACS manpower positions required to support a base with those characteristics. The shape of each base metric is the ratio of manpower among the various ACS career fields. For example, an MCO scenario with a very dense beddown (i.e., many aircraft at few locations) would, on average, have a lower ratio of firefighting manpower to services manpower than would a scenario with a more-dispersed beddown. Additional firefighting capability is needed to open each new base; therefore, twice the number of bases would require roughly twice the firefighting requirement. Services' manpower requirements are driven primarily by the base population. A more-dispersed beddown would require only marginally more total personnel and, consequently, only marginally more services personnel. Thus, as equal numbers of aircraft are spread out at more bases, firefighting requirements increase proportionally, while services' requirements increase at a slower rate. Hence, the ratio among career-field populations changes as the beddown changes.

Once aware of the characteristics of surge and steady-state operations, we can then reshape ACS forces to support them.

Rebalance Active-Duty, Reserve-Component, and Civilian Forces

The third component of our analysis was to rebalance each of the three pools of manpower: AD, ARC, and civilian. While the problem of sizing and shaping USAF manpower is compatible with optimization solutions, we did not take an optimization approach. We instead chose to use a few simple steps to reshape ACS manpower such that the AD and ARC can provide capabilities that are more balanced to the expeditionary demands described above. Later in this chapter, we discuss alternative, optimization-based approaches.

As Figure 2.1 shows, we rebalanced each pool of manpower separately. When we say *rebalanced*, we mean that we adjusted the size of each career field (within each pool of manpower) such that the new force mix is balanced against a given objective. First, we rebalanced the AD to align with steady-state operations. In other words, we adjusted each ACS career field (within our scope) in the AD such that the AD could produce a whole number of steady-state bases, with no excess capacity in any career field. Thus, the AD was balanced against the steady-state metric we established above.

While the ARC has extensively supported steady-state operations in recent years, we chose to rebalance only the AD to support steady-state operations for two reasons. First, the active duty can produce more steady-state capacity per person because of the difference between active and reserve deploy-to-dwell (D2D) ratios (1:2 and 1:5, respectively, according to USAF policy). Thus, rebalancing this way provides the most "bang for the buck" from the standpoint of end strength. We do not exclude the ARC from steady-state operations; we merely shape the AD so it produces the maximum capacity given a fixed end strength.

The second reason we rebalanced in this way is because the ARC support of steady-state deployments depends on volunteers, and we cannot assume that the ARC will *always* have volunteers available in all career fields at the rates that were planned and expected. If we were to assume ARC forces were available at, say, a 1:5 D2D ratio at all times, that could lead the USAF to divest itself of some AD capability that the ARC would be expected to supply. If, when a contingency occurred, the ARC was not able to provide those forces, a greater burden would fall to the AD force that had been reduced. We simply take a conservative approach and size the AD to meet the full extent of anticipated steady-state commitments, leaving the ARC as a reserve to primarily support surge operations. Any ARC forces available and needed to meet steady-state demands can be used to do so.[32]

Now we explain how we rebalanced active-duty manpower to meet steady-state operations. Our steady-state base metric defines the ratios among ACS career fields that match anticipated steady-state demands. We seek an AD force whose manpower has the same balance among career fields, such that the AD can always provide equal increments of steady-state capacity. If, for example, the AD lacks firefighting capacity relative to SF capacity, the USAF will likely not be able to support expeditionary bases once it exhausts its firefighting capacity. More SF capacity, at that point, will do little good. All other things being equal, that expeditionary SF

---

[32] DoD guidance presents two separate but compatible views on the subject of the utilization of the reserve component in steady-state operations. Guidance on force sizing and shaping explicitly states that active duty forces should be sized and shaped to support rotational operations. See United States Department of Defense, *Defense Planning Scenario: Steady-State Security Postures/Integrated Security Postures (SSSP/ISP)*, Final, Washington, D.C., April 18, 2008a, not available to the general public. On the other hand, the reserve component was used extensively in Operation Iraqi Freedom and Operation Enduring Freedom. Thus, DoD also published guidance outlining permissible D2D ratios for the reserve component in rotational operations. See United States Department of Defense, *Managing the Reserve Components as an Operational Force, Department of Defense Directive 1200.17*, Washington, D.C., October 29, 2008b, and also DoD, 2008a. We see the former as DoD's idea of how the force ought to be shaped and the latter as DoD's instruction on how far the services may go in utilizing reserve forces when the active component has insufficient forces to meet rotational demands (i.e., how to *manage* the reserve component). With that in mind, our method of reshaping the force aligns with DoD guidance on force sizing, but we still include reserve forces in our calculations of rotational capacity, to show how much capacity would be accessible should the reserve component be used in rotational operations.

capacity is unusable and would be better allocated to a functional capability, like firefighting, that would actually produce expeditionary capacity.

Thus, we took the supply of manpower and shifted manpower positions from career fields with relatively more expeditionary steady-state capacity to those with relatively less. This produced an active-duty force that was balanced purely to support those steady-state operations as efficiently as possible.

We then took that rebalanced active-duty force and assessed its capacity to support surge bases. We added to that the current ARC manpower mix, and balanced the ARC (i.e., shifted manpower positions within the ARC as we had for AD) such that the total AD and ARC force provided a balanced capability of surge bases. Thus, the ARC force was reshaped to complement the reshaped AD such that the *total force* provided balanced surge capacity.

The final step was to rebalance the civilian population to support home-station operations. When we say *home-station*, we refer to day-to-day installation support at bases with permanently stationed forces, where those forces perform the majority of their training activities. This excludes employ-in-place forces, such as space and missile operators, and personnel who perform institutional activities, such as headquarters staffs, which we take as a given at their current force mix.

We assume that the current AD, civilian, and contractor populations are sized to support home-station operations, such that the total of those populations at any given base reflects the total personnel needed to support that base's home-station operations (notwithstanding any shortfall against validated requirements). The active and civilian authorizations are reflected in unit manning documents (UMDs), and that is where we focus.[33] When we rebalanced the AD force according to expeditionary demands (in our first step), this left the total home-station force imbalanced. Thus, where a career field lost AD positions, it had too few total manpower positions to support the home station. In those cases, we added civilian positions to make up for any AD losses. However, where a career field gained AD positions, we removed civilian positions to rebalance according to the UMD total manning level.[34]

Ultimately, the active-duty and civilian force in each career field returned to its original total level, providing the same total level of home-station support as before, but with an altered military-civilian balance (hence the arrow from the "Rebalance AD" box to the "Rebalance civilian" box in Figure 2.2). This is not what is referred to as *backfill* (i.e., the use of civilians, contractors, or reservists to *temporarily* fill a hole in a unit left by a deployment). What we are doing represents a *permanent* shifting of manpower positions from one career field to another.

---

[33] Existing Air Force manpower data do have some estimate of contractor positions, but they are not always kept up to date.

[34] These active duty-to-civilian substitutions were always subject to logical constraints, such as the actual numbers of active duty and civilians in each career field. We also disallowed the substitution of civilians in career fields where civilians were few (e.g., many medical service career fields) or essentially nonexistent (e.g., EOD and emergency services).

18

In reality, if the USAF planned to make substantial active-duty reductions in a career field that would result in a potential gap in home-station support, there are at least three options for addressing the personnel shortfalls. The first, as we discuss here, is to increase civilian manpower positions to redress the imbalance. Second, for some installation services, the USAF could seek contractor support. Many of the installation-support activities the USAF provides through military or civilian manpower are provided by contractors in the other military services and in private industry. Finally, for a limited set of base activities, the USAF could choose to end support in those areas, taking the active-duty reductions as pure savings.[35]

By realigning according to the steps outlined above, we balanced (and therefore increased) expeditionary capacity. These realignments were subject to the following constraints:

- **Meet current funded home-station authorizations.** As described above, we took current UMD manning levels as a fixed requirement and balanced the civilian workforce such that the sum of AD and civilians met this requirement. It is the case that some career fields have rather extensive manning shortfalls, i.e., validated but unfunded manpower requirements. There would thus be a gap between the validated requirement and the home-station requirement to which we rebalance. We discuss this challenge in Chapter Four.
- **Non-deployable positions are a fixed cost**. This includes institutional and employ-in-place positions. In the same way that we excluded these from the supply of forces for our capacity calculations, we excluded these positions from the available pool for realignment. Further analysis is necessary to address reshaping these positions to better meet institutional needs and strategic objectives.
- **Fixed end strength.** We fixed our total resource pool of manpower in each component separately. Our first alternative manpower mix uses the current end strength of the AD and ARC as a ceiling. We also explored alternatives using the same method, but with reduced AD and ARC end strength, which we explain below.
- **Maintain the skill balance.** Because the skill mix for expeditionary deployment packages does not always align with the overall balance within a career field (e.g., deployments often have higher proportions of senior enlisted), we ensured each career field's skill mix mirrored its authorized mix. This helps ensure career progression can be maintained in a reduced or increased manpower pool.

Our Approach Has Two Important Characteristics

There are several implications of following the steps outlined above. First, we integrated expeditionary and home-station requirements rather than favoring one or the other. This works within the current end-strength constraints to best meet both competing demands.

At the same time, by balancing across career fields, we took an enterprise view of ACS. This ensures the capability of each career field in the expeditionary context is not out of balance with other career fields. It also ensures these ACS capabilities are in better alignment with operational capabilities.

---

[35] Recently, the USAF has explored these options in an effort called Global Basing Strategy.

Our Approach Does Not Include All ACS Capabilities

By reducing expeditionary ACS demands to a metric of bases, we naturally exclude some important ACS capabilities that cannot be so easily simplified. First, we exclude theater-level assets that depend on more theater-level inputs (e.g., RED HORSE, aeromedical evacuation hubs, and large hospitals).[36] Second, we exclude UTCs for a number of rarely used capabilities that the USAF still needs to maintain (e.g., deployable depot repair teams). Finally, we exclude response forces such as contingency response groups. Those units are subject to other demands and constraints. For those capabilities outside our scope, we advocate using separate methods to calculate requirements. A comprehensive assessment of ACS manpower requirements would apply our base-driven approach to as many capabilities as possible, separately analyze the excluded capabilities we list above, then integrate all these requirements into a single, coherent picture so that planners and decisionmakers can understand the requirements and shortfalls across the force.

Our Approach Does Not Fully Address Home-Station Demands

Our approach uses as a baseline the current levels of UMD manning for home-station operations. In a surge scenario (presuming full deployment), this leaves civilians, contractors, non-deployable positions ("DX coded"), and 20 percent of the deployable active-duty forces (many of which would be fit for home-station tasks) at home stations. In a steady-state scenario, our approach would leave all of the above forces plus two-thirds of the active duty force (given a 1:2 D2D ratio) at home stations.

The problem is that this does not guarantee an adequate level of support at home stations (which may be different under different deployment conditions and at different bases). During a deployment of active-duty forces, which creates a net shortfall in home-station personnel, a local commander could mitigate the shortfall by having people work more hours, cutting services, funding backfill (via civilians, contractors, or reservists), temporarily reallocating certain personnel to more-critical tasks, or some combination of the four.

The question of how many people are "really needed" at home stations during a contingency does not have a straightforward answer. Other PAF research addresses this topic in detail.[37] In this report, we limit the D2D ratio of the active duty to 1:2, thereby limiting the extent of home-station shortfalls in any career field. Reshaping the force to provide more-balanced expeditionary capabilities ought to also balance the impact of home-station shortfalls more evenly among ACS career fields.

---

[36] In some cases, these theater-level capabilities utilize personnel in the same AFSC as the base-level capabilities we do include. One example would be RED HORSE, which is composed of personnel from the same AFSCs as engineering operations. We avoided such a conflict by excluding personnel at the UTC. For example, we excluded RED HORSE UTCs altogether, essentially setting aside the manpower positioned in those UTCs.

[37] Patrick Mills, John Drew, Dan Romano, John Ausink, Mike McGee, and Therese Bohusch, *Beyond Deploy-to-Dwell Ratios: New Metrics to Inform Home-Station Capability*, unpublished RAND research.

Our Approach Is Not an Optimization

As mentioned earlier in this section, we do not perform an optimization (e.g., linear program [LP]) to rebalance ACS manpower. It would be possible, however, to apply such techniques in a way that is compatible with our overall approach of using the metric of expeditionary bases. One could formulate a linear program to maximize two goals, making one a predetermined priority.[38] One could perform an optimization such that either steady-state or surge bases were a priority. With both solutions in hand, one could work out the trade-offs between surge and steady-state bases and possibly find some combinations that support more of both types of bases than the solutions we reach through a sequential, arithmetic approach.

Conclusion

The goal of the analysis in this report is to not to find the "optimal" solution to the question of ACS force structure, but to illustrate very simply how much can be gained by reshaping toward expeditionary objectives. Using the steps described in this chapter, we lay out an approach to manpower shaping that integrates expeditionary and home-station requirements, balances among functional areas based on a single set of objectives, and articulates capacity in a way that is readily translatable to operational objectives. In Chapter Three, we demonstrate this methodology quantitatively.

---

[38] One way to do this would be to construct a variable that equals steady-state bases supported times 1,000, plus surge bases supported. Maximizing that variable has the effect of maximizing both, but with a priority on steady-state bases. Constructing the variable in the opposite direction (multiplying surge bases by 1,000) would prioritize surge bases.

3. Current and Alternative ACS Manpower Mixes

In Chapter Two, we explained our approach to developing ACS manpower requirements: reshaping ACS forces to meet sets of steady-state and surge demands while supporting home-station operations. In this chapter, we demonstrate that approach quantitatively. We first show the capacity of the current ACS forces to support future surge and steady-state operations. Then we show the results of realigning manpower among active, reserve, and civilian workforces, and we conclude with a discussion of the implications.

Capacity of Current ACS Manpower

Assumptions for Capacity Calculations

We make the following assumptions in our calculations. A more detailed account of our methods and assumptions is contained in previous unpublished RAND research.

- The supply of forces (i.e., quantities of manpower in each career field) is determined in accordance with current data from the UTA file and AFI 10-401.[39]
- Eighty percent of manpower is available at the time of deployment.[40]
- Only USAF forces are supported. This is sometimes referred to as "blue on blue" support.
- Joint expeditionary tasking (JET) demands are not included.
- The capacity of a career field is determined by the limiting AFSC.
- AD and ARC are accessed according to current USAF policy, a 1:2 D2D ratio for the AD and a 1:5 D2D ratio for the ARC.[41, 42]

Expeditionary Capacity

Figure 3.1 shows the surge capacity of the career fields in our scope. We measure this by simply dividing the total manpower available in each career field by the total manpower required for a single surge base, following the methods laid out in Chapter Two.

---

[39] Air Force Personnel Center, UTA file, 2011; AFI 10-401, 2005.

[40] According to United States Air Force, 2011a.

[41] Norton Schwartz, "Air Force Strategic Choices and Budget Priorities Brief at the Pentagon," Washington, D.C., January 27, 2012; and Air Force Instruction 10-402, *Mobilization Planning*, Washington, D.C., May 1, 2012.

[42] The D2D ratio at which a given career field operates has near- and long-term impacts on career field health and often on home-station support. We neither advocate for the current policy for D2D ratio nor any alternative D2D ratio. We merely use this to show the net result of following current policy. We later discuss the implications of active duty deployments on home-station operations.

Figure 3.1 Total Force Expeditionary ACS Surge Capacity

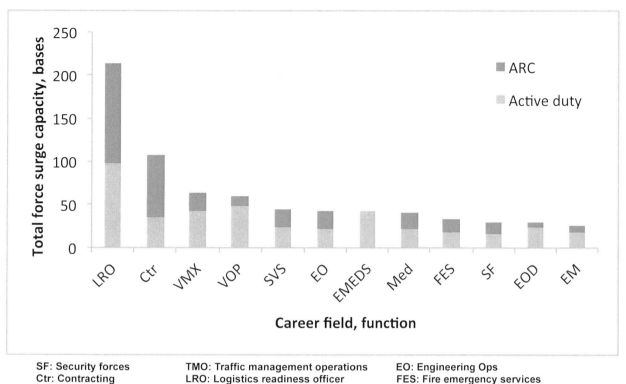

SF: Security forces   TMO: Traffic management operations   EO: Engineering Ops
Ctr: Contracting   LRO: Logistics readiness officer   FES: Fire emergency services
VMX: Vehicle maintenance   EM: Emergency management   EOD: Explosives Ordnance Disposal
VOP: Vehicle operations   EMEDS: Expeditionary medical support SVS: Services

In Figure 3.1, the height of each column shows the number of surge bases the *total force* can support, expressed by the capacity of the limiting resource (i.e., AFSC) in each career field. The colors in each column represent the capacity in the active duty (blue) and the reserve (red). It can be seen that the current ACS manpower mix is not balanced against the general deployment characteristics of long-range surge defense planning scenarios (the primary source from which we derived our surge-base metric). This is to be expected. The current ACS force is balanced primarily based on home-station requirements (along with their civilian counterparts) and career-field needs such as career progression.

From a long-term manpower planning perspective, then, the USAF can only provide expeditionary base support up to the level of its limiting resource—in this case, emergency management. The capacity of EM, according to these calculations, is 23 surge bases. Depending on the expeditionary requirements, this may be enough. Whatever the "real" requirement for surge bases to beddown USAF forces, this imbalance translates to either great excesses of capacity (if the requirement is lower), or great deficits (if the requirement is higher), or both. We now contrast this with an assessment of steady-state capacity.

Figure 3.2 shows the steady-state capacity of the career fields in our scope, again for the total force. In this figure, the career fields appear in the same order as in the previous figure. Here, the blue portion of the column reflects the AD capacity at a 1:2 D2D ratio, while the red portion reflects the total force capacity with the ARC at a 1:5 D2D ratio. For each, we divide their capacity to support bases by the appropriate D2D divisor: three and six, respectively.

Figure 3.2 Expeditionary ACS Steady-State Capacity

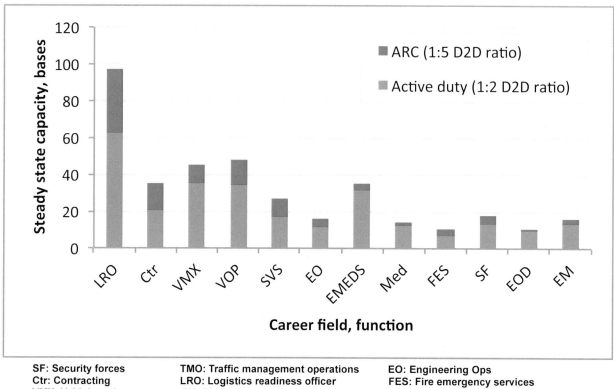

SF: Security forces	TMO: Traffic management operations	EO: Engineering Ops
Ctr: Contracting	LRO: Logistics readiness officer	FES: Fire emergency services
VMX: Vehicle maintenance	EM: Emergency management	EOD: Explosives Ordnance Disposal
VOP: Vehicle operations	EMEDS: Expeditionary medical support	SVS: Services

The difference in the relative heights of the columns in Figures 3.1 and 3.2 stems from two sources. Each career field has a slightly different balance between the components, so when different D2D ratios are included, the total capacity varies. The other reason for the difference is that steady-state and surge bases do not require the same mix of support. In the base metrics we developed, for example, steady-state bases require very little EM (the rightmost column). In this case, a comparable number of EM personnel divided by a smaller number (steady-state bases require less than surge bases) results in a higher relative capacity. The height of each column reflects some degree of both of these factors.

Now that we have assessed the capacity of the current force to support surge and steady-state operations, we will display this in a more-condensed format using a new graph.

In this graphic, we envision a career field as a pool of resources that can be "spent" on expeditionary operations until the resources run out. The USAF can spend these resources on surge operations (in which every available resource is sent to deploy), steady-state operations (rotational operations in which part of the deployable force is at home stations preparing to deploy or recovering from deployment), or a combination of the two (if, for example, a steady-state operation is ongoing, and the requirement for a surge operation arises).[43] When the USAF has spent (i.e., deployed) its last resource, it has reached the limit of its expeditionary capacity. This analysis focuses on generic ACS capabilities that would deploy to most any expeditionary base.

Figure 3.3 illustrates this approach. On the left side of the figure, we present a notional force from a single functional area with 1,800 manpower authorizations, all of which are available for deployment, and all of which contribute equally to expeditionary operations (i.e., independent of skill level), as represented by the black column on the left. We then assume that all expeditionary bases—both surge and steady-state—require 100 personnel from this career field. Given this assumption, the surge capacity of that career field can be represented as 1,800/100 = 18 surge bases to support the beddown and operation of aircraft. This is shown by the blue column labeled Case 1 on the left-hand side of Figure 3.3. In this case, all forces are supporting surge bases.

Figure 3.3 Notional Example of Expeditionary Manpower Capacity

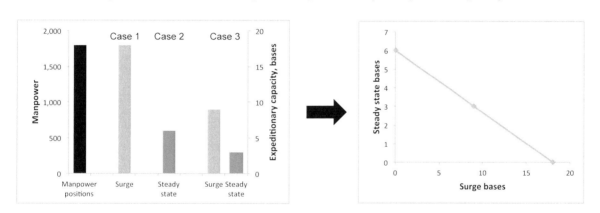

We can translate this metric into operational terms by thinking specifically of MCOs, the type and number of aircraft they require, and the threats to which the bases are vulnerable. If operational planners require 15 bases to support the beddown of, say, 1,000 aircraft, our notional career field can support that deployment. If, however, planners estimate that they will need to disperse their forces to 27 bases to diffuse enemy attacks, this career field would either be unable to support that beddown, or it would be forced to support 27 bases with 18 bases worth of capacity, or 18/27 = 67 percent of the career field at each base (or some variation thereof). Thus, this career field can use the metric to quickly communicate its capacity to support surge operations in a way that can inform operational planning decisions.

Moving to the red column labeled Case 2 in Figure 3.3, we calculate steady-state capacity in the same way, except that we now include rotations. With the same pool of 1,800 personnel, and

---

[43] This last case is not unlike recent operations, in which forces were conducting steady-state, or rotational, operations in Afghanistan, then conducted surge operations in Iraq, and then transitioned to steady-state operations in Iraq.

assuming a D2D ratio of 1:2 (assuming active-duty support only), these forces can support 1,800/100/3 = 6 steady-state bases on a sustained basis without violating that D2D ratio. The height of the bar represents a capacity of six steady-state bases. The remaining forces not supporting steady-state bases at the moment are at home stations recovering from deployment or preparing to deploy.

To visualize that in operational terms, a large COIN operation might take five bases (for USAF forces), a disaster-relief operation might take only one base, and a small show of force might take three bases to beddown USAF aircraft and personnel. Hence, that pool of personnel could support either the COIN operation and the disaster-relief operation or the show of force and the disaster-relief operation, but not all three operations simultaneously.

Thus, our pool of personnel from this career field can support 18 surge bases or 6 steady-state bases, but not both simultaneously. If the USAF sought to operate surge and steady-state bases simultaneously, the force could be split in two, for example, to support 9 surge bases and 3 rotational steady-state bases.[44]

We now plot these base combinations—18 surge and 0 steady-state, 0 surge and 6 steady-state, and 9 surge and 3 steady-state—on the graph on the right side of Figure 3.3. In this graph, the x-axis shows the capacity to support surge bases and the y-axis shows the capacity to support steady-state bases. This career field can support deployments that sit anywhere on the line or below it (i.e., less), but not any more than that. That is the maximum capacity of the career field.

This display, then, provides a quick look at the expeditionary ACS capacity of any set of resources, in this case manpower in a career field. We now conduct this same assessment for the career fields in our scope using our actual data from Figures 3.1 and 3.2.

Current Expeditionary ACS Capacity Is Limited by ACS Imbalances

In Figure 3.4, the x- and y-axes are the same as in Figure 3.3. Here, we include the total force for steady-state operations, drawing from the data in Figure 3.2. We have plotted 11 of the 12 ACS career fields (or groups thereof) included in this assessment: SF, services, EOD, fire protection, EM (which handles CBRN threats), engineering operations (aka PRIME BEEF), contracting, fuels support, EMEDS, traffic management, vehicle operations, and vehicle maintenance (based on AFSC-level calculations).[45] Each gray line represents the capacity of an individual career field. We have highlighted in black the capacity of the limiting resources at the

---

[44] This combination of surge and steady state assumes the surge bases are supported until the end of the operation.

[45] The curve for logistics readiness was far to the right of the other curves, so we truncated the display for easier legibility.

Figure 3.4 Current Expeditionary Capacity of Selected ACS Career Fields

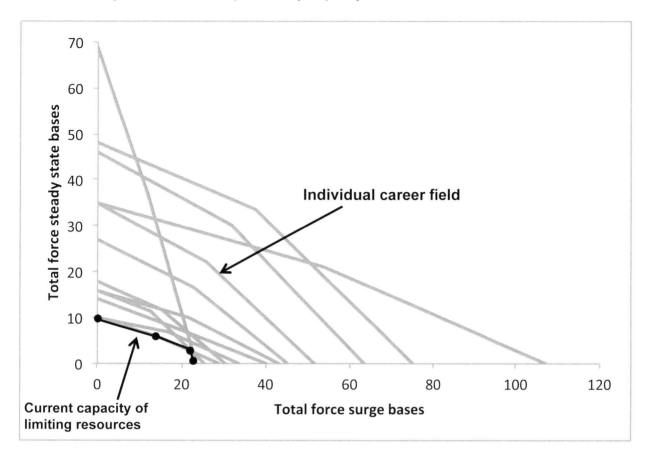

bottom left of the graph. This line is defined by the intersection of three career fields: (1) EM for surge bases, (2) firefighting emergency services (FES) for steady-state bases, and (3) EOD for the line connecting them.

We now walk through an example career field to explain how best to read this graph. The line representing transportation management operations (TMO) is indicated by an arrow in Figure 3.4. If we look at the y-axis, we see that TMO intersects the axis at about 35 steady-state bases. This means that, given the manpower postured in the AD and ARC, TMO could simultaneously support 35 steady-state bases according to the parameters laid out above. If, while the force sustained those operations, a surge contingency arose, forces would be diverted from steady-state operations to support the surge. Moving along the TMO line to the right, it can be seen that the current force could support 26 surge bases while reducing the steady-state capacity to 21 bases (shown by the kink in the line). For this computation, we assume the ARC is the first to leave the steady-state operations to respond to a surge because, on a per-person basis, it offers equal surge capacity but half the steady-state capacity. In reality, AD forces would likely be the first to respond to a surge operation (due to mobilization and timing constraints). The graph is meant to illustrate the outer envelope of capacity in each area.

If still more surge capacity was needed, more forces could be diverted from their rotational operations. Continuing down the line to the right, the USAF could provide additional surge capacity up to a total force capacity of 52 surge bases, which would require vacating all deployed steady-state bases.

We make several observations regarding this figure. First, as with the assessments from Figures 3.1 and 3.2, these lines, which represent individual career-field capacity, are imbalanced relative to these scenario types. The capacity to support surge bases ranges from 25 bases to over 200. To reiterate, these assessments are derived by dividing the number of available manpower positions by the number of positions in each AFSC required to support each base type. There is some clustering in the lower ranges, but many career fields are spread throughout the range. The case is similar for steady state. The capacity of these career fields ranges from about ten bases up to nearly 100.

To the degree that these two metrics—surge and steady-state bases—capture the range of future deployment demands (and we have synthesized scenario requirements to achieve that end), the USAF would get the most-effective and efficient force if all the lines on the graph were in balance with one another, essentially overlapping. In this case, SF and EOD would be able to support the same expeditionary demands, as would firefighting and logistics readiness officers, and so on. Because we assess that these metrics do approximate the operational dynamics of the scenarios on which we focus, we seek such a balance.

Manpower Mix Options

Assumptions and Data Sources for Realigning Manpower

We make the following assumptions when realigning manpower across career fields:

- ARC personnel postured as available are not needed for home-station installation responsibilities. This assumes ARC-only bases use Air Reserve Technicians (ARTs), Air Guard Reserves (AGRs), and civilians for day-to-day support.
- For home-station activities, workload could be assigned to either an AD or civilian workforce. For the purposes of providing base support and training junior personnel, civilians can accomplish the same tasks as active-duty personnel at a comparable level of productivity.
- To estimate annual cost savings of manpower realignments and reductions, we use fiscal year 2012 Standard Composite Rates.[46]

Rebalancing Capacity Entails Major Manpower Shifts

Following the steps outlined in Chapter Two, we calculated the manpower shifts necessary to balance expeditionary capacity for AD and ARC to support steady-state and surge operations. To

---

[46] AFI 65-503, 1994.

do so, we used a Microsoft Excel spreadsheet tool developed for this purpose. Our tool catalogs each career field's available manpower, home-station manpower requirements, authorized skill mix, and manpower costs.

Figure 3.5 shows the specific manpower movements among career fields and capability areas if current USAF end strengths are available. In the figure, career fields and capability areas are shown along the x-axis, while net manpower additions and reductions are shown on the y-axis. AD is shown in blue; ARC is shown in red; civilians are shown in green.

Figure 3.5 Manpower Realignments by Career Field

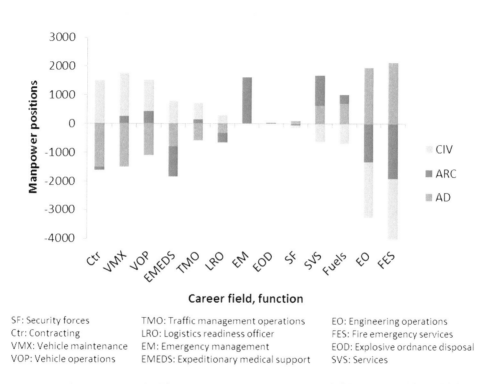

SF: Security forces	TMO: Traffic management operations	EO: Engineering operations
Ctr: Contracting	LRO: Logistics readiness officer	FES: Fire emergency services
VMX: Vehicle maintenance	EM: Emergency management	EOD: Explosive ordnance disposal
VOP: Vehicle operations	EMEDS: Expeditionary medical support	SVS: Services

On the left side of Figure 3.5, contracting shows the realignments clearly. Contracting has, according to these metrics, relatively more steady-state capacity in AD than do other career fields. Thus, we reduced the AD by about 1,400 positions and added equivalent civilian positions to offset the home-station loss. Bear in mind, when we realign manpower in this way, it ensures a balanced AD D2D ratio across the force. Further, the addition of civilians makes home-station support more reliable, since civilians will not generally deploy.

As shown in Figure 3.1, EM had the lowest relative surge capacity, so we added about 1,600 positions to the ARC. Because, as seen in Figure 3.2, the steady-state demand for EM is relatively low, the capacity could be added entirely to the ARC, as it is a less-costly means of providing surge support. The AD was not cut, even though its capacity exceeded the steady-state capacity of most other career fields. The reason for this is that we assume the positions are needed for home-station operations. Because civilians do not generally support EM (thus precluding an AD-civilian substitution), we left the AD force as it was. In most other cases, a steady-state excess would have resulted in an active-duty reduction.

FES had the lowest relative steady-state capacity, so thousands of positions were added to the AD force. Then, ARC positions were reduced because the total force surge capacity increased. Finally, civilians were reduced to level out home-station capability. The net result of all these shifts is a force rebalanced to meet expeditionary and home-station objectives. In the next section, we show the results of these manpower shifts in terms of additional expeditionary capacity.

USAF Can Achieve More Expeditionary Capacity with the Same End Strength by Realigning Manpower

Figure 3.6 shows the capacity of the current manpower mix and three alternative manpower mixes. The arrangement of this figure is the same as that of Figure 3.4, with the x-axis showing the capacity to support surge bases and the y-axis showing the capacity to support steady-state bases. The x- and y-axes have been truncated for easier viewing.

The first alternative mix, labeled Maximum, balances manpower according to the steps outlined in Chapter Two, staying within the current AD and ARC end strengths (as defined by the manpower authorizations in each career field in the UTA file). This shows the maximum expeditionary capacity the USAF could achieve without increasing end strength. The second and third alternatives simply decrease the expeditionary capacity of the first alternative by 25 percent and 50 percent, respectively, to achieve end-strength and budget savings.

Figure 3.6 Manpower Mix Options

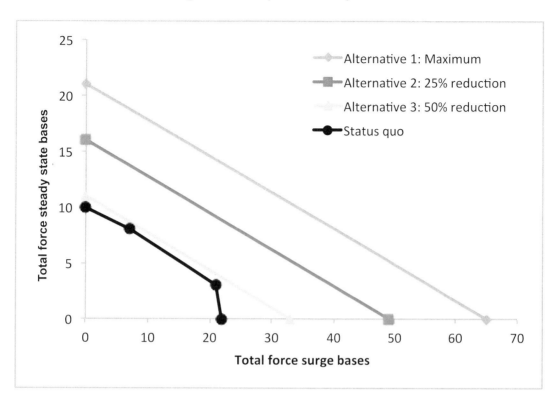

Figure 3.6 shows only the capacity of the limiting resource for each manpower mix. The black line, labeled "status quo," shows the minimum from Figure 3.4. The blue line, labeled Maximum, shows the capacity of the first alternative manpower mix. These realignments take manpower from career fields with more expeditionary capacity and move it to those career fields with the lowest capacity until all career fields reach equilibrium relative to our expeditionary metrics. Alternative 1 results in an expeditionary force that can support 65 surge bases with the total force, 21 steady-state bases with the AD, or any combination of both that lies at some point along the blue line.

Alternatives 2 and 3 show the expeditionary capacity that results from cuts to the total force that result in a 25 percent and 50 percent reduction in expeditionary capacity, respectively. The relative proportions are defined by the slope of each line and are the result of the relative size of the AD and ARC components. These alternatives still perform better than the limiting resources of the current ACS force, with Alternative 3 coming very close to today's limiting resources.

We make one additional point here before moving on. The capacity lines in Figure 3.6 are straight rather than curved, or kinked, like the lines in Figure 3.4. Those kinks exist because the AD and ARC contribute different amounts to steady-state operations. Looking at the TMO example in Figure 3.4, if one starts from the point where all forces are engaged in surge operations, 52 bases are being supported. If one diverts AD forces from surge to steady state, those forces get a lot of "bang for the buck" because of their 1:2 D2D ratio. Once AD forces run

31

out and ARC forces are diverted, each ARC surge base dismantled provides only half the steady-state capacity as the AD due to D2D ratio limitations.

In our rebalanced force, some career fields have zero ARC capacity. The algorithm allocates manpower to career fields such that surge and steady-state capacity are both balanced. In some cases, the action that maximizes *total force* capacity is diverting all ARC manpower from one career field to another. This leaves some career fields with a robust AD force but no ARC forces.

It is certainly possible to formulate the rebalancing differently. One could employ constraints that require every career field to have some AD and ARC capacity or that require certain career fields to have ARC capacity (e.g., those career fields with emergency-response roles to provide Air National Guard capabilities to states). Such constraints would reduce total force capacity in the interest of pursuing some competing objective.

While some career fields have zero ARC capacity in our rebalanced force, most have some. These career fields would thus have some steady-state capacity that is above the line of the limiting resource shown in Figure 3.6. This capacity would be in addition to the balanced steady-state capacity shown in the figure and could be utilized if volunteers were available.[47]

As stated earlier in this section, Alternatives 2 and 3 entail significant manpower reductions and would bring expected savings. We translated these manpower shifts into recurring cost savings, shown in Table 3.1.

Table 3.1 Recurring Savings Associated with Manpower Realignments

Alternative	Manpower Positions			Annual Savings	Total Positions Changed
	AD	ARC	Civilian		
Alternative 1: Maximum	-250	-850	250	-$1M	37,500
Alternative 2: 25% reduction	-9,700	-6,200	9,700	-$340M	39,200
Alternative 3: 50% reduction	-19,200	-11,600	19,200	-$670M	52,400

NOTE: Numbers are rounded.

In this table, the first column shows the three alternatives discussed above. The second, third, and fourth columns show the net increases or decreases in active, reserve, and civilian manpower positions, respectively, resulting from each alternative manpower mix. The fifth column shows the net annual cost or savings that would result from those manpower changes. To calculate the cost and savings, we used 2012 Standard Composite Rates from AFI 65-503.[48] For the sum of

---

[47] It might be that sacrificing a little surge capacity could produce enough steady-state capacity to be worth the trade. Exploring such trade-off questions is possible with an optimization approach.

[48] AFI 65-503, 1994.

manpower realignments in each alternative, we estimated the cost or savings associated with gaining or losing each position, using applicable rates for each component.[49] This reflects only the annual recurring costs and does not include any investment that would be necessary to make these realignments on some implementation schedule.[50] Because the Defense Health Program (DHP) funds most medical manpower positions, we excluded medical positions from our calculations of potential savings from reduced end strength.

The last column in Table 3.1 shows the total manpower positions changed for each alternative. This includes all additions or reductions in each career field or component. For example, a shift of one position from one career field to another in the active duty equals four changes: an active duty reduction from the first career field, an active duty addition to the second, and a corresponding addition and reduction in civilian positions to support the home station.

For Alternative 1, the costs do not exactly break even. We chose to balance AD and ARC manpower to equal a whole number of bases so that the USAF would not be left with a force that was exactly neutral in terms of manpower or dollars but provided unusable expeditionary capacity. Thus, we used the current AD and ARC end strengths as ceilings and balanced their capacity to produce a whole number of expeditionary bases without exceeding current force levels. The savings generated by the small cut in AD forces is offset by the addition of civilian positions to support home-station operations, and there is a nontrivial reduction in ARC forces to result in a balanced expeditionary capability. The total recurring savings for the USAF would be about $5 million per year.

Alternative 2 accepts a reduction of about 9,700 AD and 6,200 ARC manpower positions and adds 9,700 civilians to maintain support for home-station operations, resulting in an estimated annual savings of $340 million per year from reduced manpower. Alternative 3 takes those cuts further with a reduction in manpower of 19,200 AD positions and 11,600 ARC positions. To substitute for lost AD manpower, 19,200 civilian positions were added. This would result in an annual savings of roughly $670 million.

Rebalanced ACS Posture Better Meets Expeditionary Scenario Demands

Up to this point, we have focused on the relative balance of AD and ARC forces to meet generic steady-state and surge demands (as articulated in the range of defense planning scenarios) and the trade-offs of cost and capability. We now take up the question of how well these alternatives perform against specific scenario sets. The following section is meant to

---

[49] We estimated the costs and savings using grade-specific and non–grade-specific costs, but specifying the grade made very little difference.

[50] For example, reducing a career field by one thousand positions would probably take years to let some of separations happen naturally through attrition and retirements. Likewise, recruiting and training many more people for a career field would require not just an investment of money, but also a delay as adequate experience was built up to have competent personnel.

illustrate how a rebalanced force can be calibrated against specific scenarios, not to advocate for any particular set of planning scenarios or guidance document.

As discussed in Chapter One, OSD develops force-sizing constructs comprising overlapping sets of planning scenarios. We use example scenario sets similar to those found in the 2010 QDR. The QDR lays out three sets of scenarios against which the military services are instructed to size and shape themselves.[51] These scenarios are simply referred to as Sets A, B, and C:

- **Set A:** A major stabilization operation, deterring and defeating a highly capable regional aggressor, extending support to civil authorities in response to a catastrophic event in the United States
- **Set B:** Deterring and defeating two regional aggressors while maintaining a heightened alert posture for U.S. forces in and around the United States
- **Set C:** A major stabilization operation, a long-duration deterrence operation in a separate theater, a medium-sized COIN mission, and extended support to civil authorities in the United States.

We use these scenario sets as a jumping-off point to perform some illustrative calculations. We slightly adapt these scenario sets into the following three cases:

- **Case A:** A medium-sized rotational operation consisting of five steady-state bases and a single, stressing major combat operation with an extremely dispersed beddown consisting of 40 surge bases.
- **Case B:** Two simultaneous surge operations with moderately dispersed beddowns, each consisting of 25 surge bases.
- **Case C:** Many small, dispersed rotational operations (e.g., regional stabilization, no-fly zone, show of force), totaling 12 steady-state bases.

Figure 3.7 shows the expeditionary capacity of current and alternative manpower mixes against these three cases. The figure includes the same axes and capacity lines as Figure 3.6; however, we have added one dot for each case to show the approximate size of its demand: blue for Case A, green for Case B, and yellow for Case C.

---

[51] United States Department of Defense, 2010a.

Figure 3.7 Expeditionary Capacity of Current and Alternative Manpower Mixes Against Illustrative Scenario Cases

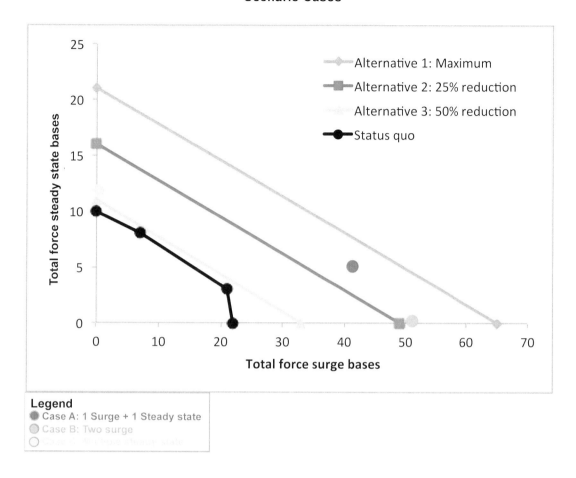

The first conclusion we draw regarding Figure 3.7 is that the current ACS manpower mix, if defined by its limiting resources, does not have the capacity to meet any of our illustrative cases with all ACS career fields. Reflecting back on Figure 3.4, we note that many of these career fields, though not all, have the capacity to support the number of bases required in our illustrative cases. Certainly in any real-world scenario, commanders must make decisions about risk and resource trade-offs and may employ forces knowing they do not have all the support they desire. But for long-term manpower planning, our approach enables decisionmakers to see the capacity of forces and adjust them against future requirements.

Referring now to the alternative manpower mixes, we find that Alternative 1 exceeds the demands for all three cases, while Alternative 2, the 25 percent reduction force mix, falls short of Case B. Were Alternative 2 chosen, decisionmakers could decide, if faced with overlapping contingencies as in Case B, if forces would be withdrawn from the steady-state bases to meet the full demand for surge bases. Alternatively, if commanders were comfortable with the risks incurred by basing at only 35 bases, the steady-state operation could continue unperturbed. Alternative 3, the 50 percent reduction force mix, is much more austere: it comes close to meeting Case C, but falls far short of the other two. Yet the spread of these options helps to inform the question of how much is enough. Even if a 50 percent reduction was chosen for the sake of cost savings, the potential capability shortfall could be compared to the expectations of planners to assess the potential operational risks of the new force.

In fact, the USAF has many options to mitigate shortfalls of the type identified in this exercise. For example, the USAF could accept more home-station shortfalls by deploying more active-duty forces, it could also use ARC volunteers in steady-state operations, use contractors for deployed installation support (in permissive environments), or it could request the support of other services for applicable functions (e.g., base security, some engineering functions), among other options. The trade-offs illustrated in Figure 3.7, and the underlying analysis, can serve as a starting point for substantive discussions about ACS cost and capability.

Recall from Table 3.1 that each alternative comes with cost savings. In the current environment of fiscal constraints and flat or decreasing budgets, accepting some potential resource shortfalls or "risk" may be necessary to garner the end strength and dollar savings provided by a particular alternative. We do not advocate any alternative. We also note that there are other reasonable alternatives that we have not shown here for brevity's sake. For example, operational objectives might lead the USAF to a different mix of AD and ARC, and thus to a different mix of steady-state and surge capacity.

Neither do we advocate using any particular scenario set to drive manpower realignments. Part of our rationale for distilling future scenarios down to their essential characteristics was to be less sensitive to any new set of scenarios put forward by OSD. The recently released Defense Strategic Guidance, as discussed in Chapter One, arguably changes those operational objectives yet again (i.e., moves the scenario dots in Figure 3.7 to reflect different operational demands).

In reference to the graphs we have shown, we note that in constructing our generic surge and steady-state bases (shown on the x- and y-axis in Figure 3.7), we sought to capture the nature of expeditionary operations, so that any balancing creates a generally robust force. However, any new set of operational scenarios or objectives would require a new dot, or series of dots, on the chart. Those dots, as instantiations of operational objectives, may change as often as do presidential administrations or secretaries of defense, if not more often. However, the basic types of contingencies the USAF faces should change less often.

Summary

In this chapter, we first assessed the expeditionary capacity of ACS manpower in a very simple and straightforward way by distilling the characteristics of future operations down to the bare minimum in terms of the number of steady-state or surge bases. By doing this, we admittedly sacrificed some of the fidelity inherent in a more-detailed base-by-base assessment of requirements. In return, we gained explanatory power—the ability to articulate ACS capabilities to an operational audience in ways that can inform operational risk assessments and trade-offs.

Second, we outlined and demonstrated a method for realigning manpower that has attributes the current manpower system lacks. We have not incorporated all necessary details of manpower planning, such as force flow from active to reserve (or civilian) forces. We do, however, incorporate both home-station and expeditionary requirements for all the career fields in our scope, and we use a consistent set of requirements across career fields. Also, we make explicit important assumptions the USAF must grapple with to shape its forces for the future (e.g., the respective roles of the AD and ARC in steady-state and surge operations and rotational D2D ratios), and we have incorporated these uniformly in our analyses.

Third, by realigning forces to support expeditionary demands shaped by the characteristics of future operations rather than individual scenarios, our approach results in a force that is robust to changes in the specific scenarios chosen as long as they do not depart radically from currently envisioned mid- to long-term operations. This allows leadership to focus on key questions like what the nature of future operations will actually be and how to set those strategic and operational objectives.[52]

Ultimately, the key to our methodology is aligning manpower to expeditionary operations. While exploring a range of different scenario sets, we found that aligning to virtually any set of expeditionary operations would provide improvements over the USAF's current ACS manpower mix, as many of the functional areas exhibiting shortfalls in the calculations shown here experienced shortfalls in a range of scenarios.

---

[52] The potential set of future strategic and operational objectives can include joint demands such as JETs. In that case, JETs are simply one more set of demands competing for resources. Seen in the context of our methodology, JETs produce an opportunity cost: every manpower position the USAF "spends" on JETs is a manpower position not applied to some other mission set.

4. Additional Considerations for Shaping ACS Forces

While we addressed several important dimensions of the challenge of sizing and shaping ACS forces to meet future operations, we now discuss some salient issues we did not directly address in our calculations. In this chapter, we discuss three important planning assumptions—joint expeditionary taskings, support to contracting personnel and other services, and response times for major combat operations—and three more-enduring challenges to ACS manpower planning.

Planning Assumptions

Joint Expeditionary Taskings

The USAF refers to joint sourcing solutions in the joint community as *JETs*. Generally speaking, a JET occurs when the preferred service for a combatant commander (COCOM) tasking is unable to support the requirement and another service fills the tasking. Examples include SF personnel guarding prisons, transporters driving in convoys, and contracting officers filling positions in the joint contracting command.

For many ACS career fields, JETs have caused, and continue to cause, severe strain on the force. Some career fields have operated at a 1:1 D2D ratio for years, many in the face of existing manpower shortfalls or skill imbalances. It is widely recognized that supporting these deployments has had a detrimental and enduring effect on the readiness of USAF forces.[53]

JETs are, by definition, outside of doctrinal USAF mission areas and would not be officially programmed for in the USAF budget. Nonetheless, JETs should be considered in the USAF's planning and expectations for sustaining the force while supporting overseas deployments. While it is beyond the scope of this report to offer solutions, any future force must be flexible and resilient enough to support the range of military operations, given current realities, and be faithful to the roles and missions designated by Congress and DoD.

Support for Contractors and Other Service Members

As of March 2011, contractors made up 52 percent of DoD's workforce in Afghanistan and Iraq, supporting a range of activities.[54] While some of these contractors provided installations support, including to other contractors, some may have created an additional support burden on

---

[53] Discussions with personnel from AF/A5X, AF/A3/5, SAF/AQX, AF/A4/7, AF/A4L, AF/A7S, AF/A7C, AF/SG during 2011 and 2012.

[54] Moshe Schwartz and Joyprada Swain, *Department of Defense Contractors in Afghanistan and Iraq: Background and Analysis*, Washington D.C.: Congressional Research Service, CRS Report for Congress, May 13, 2011.

military forces supporting those bases. In addition, base operating support integration (BOS-I) responsibilities were assigned at many expeditionary bases to a single service (e.g., Joint Base Bagram), thus creating an additional support burden on the selected service, whether in support of other services, other U.S. agencies, or coalition partners. While we excluded support to contractors, sister services, and coalition partners from our analysis, operational planners must grapple with that possibility.

A related situation is that when the services depend so significantly on contractor support, it creates the additional burden of contract oversight. This usually comes in the form of contracting officers, but it also involves engineering contract oversight for construction projects. The USAF filled many joint demands for such expertise due to the nature of operations in Iraq and Afghanistan. It is not clear that increased contract oversight has been incorporated into either current planning factors or planning scenarios.

Response Times for Surge Operations

Finally, if a major operation develops quickly enough, it can cause a quick spike in demand for military deployments. Usually, only the active component will be available to meet those short timelines, thus placing a premium on AD forces in the capability areas that might be called for in such situations. We did not explicitly constrain AD realignments to meet a set of quick-response demands, but even the career fields that sustained significant cuts in our above calculations retain some AD capacity. Our modeling approach is transparent enough that such a constraint could be included fairly simply.

Challenges to ACS Manpower

We also see three significant challenges to ACS manpower beyond the scope of what we considered in this analysis.

Unfunded Manpower Requirements

In this analysis, we used authorized manpower positions as a starting point for the supply of military forces to support expeditionary operations. However, the USAF currently has unfunded manpower requirements of many thousands of positions. While manpower standards drive manpower requirements, not all of those requirements are funded as authorizations, which appear in a UMD. We analyzed manpower requirements data to estimate the unfunded positions in the career fields within our scope.[55]

We found that, for the 12 career fields and functional capabilities analyzed above, the manpower requirements data reflected a manpower shortfall (i.e., total unfunded requirements) in the active-duty military force of over 8,000 positions (compared to about 60,000 total

---

[55] Air Force Personnel Center/DSYD, Authorized Manpower Master File (MPW), September 30, 2011.

authorized manpower positions postured in the UTA for those same career fields). About half of these unfunded positions were in SF. Using the same cost methods and data described above, we estimated the annual cost to fill those positions to be around $650 million.[56] This is a significant sum, but those manpower shortfalls are consistent with estimates provided by functional experts.

Referring back to Table 3.1, the potential savings from Alternative 3 (realigning manpower and reducing expeditionary capacity by 50 percent) were about $650 million. This is a coincidence. This means that if Alternative 3 were implemented, it could roughly offset the existing unfunded manpower requirements. This shortfall essentially translates to an increased home-station burden. Because many of these career fields earn their manpower from home-station demands (e.g., SF), any manpower shortfall means the personnel that do actually work in that career field must work harder (i.e., longer hours) to support the validated but unfunded requirements.

In truth, the USAF has been gradually underfunding manpower since it began drawing down in the 1990s.[57] In our calculations, we leveled home-station support to UMD data, which only supports funded authorizations. The deficit left by these unfunded positions is a real manpower deficit, and the burden borne by USAF personnel is a real burden. An approach like the one we propose in this report could go some distance in closing this manpower gap by freeing up end strength and dollars, but the USAF must grapple with this problem on a very fundamental level to ensure the long-term sustainability and readiness of the force.

Sustainability of a New Active-Reserve-Civilian Manpower Mix

Most ARC personnel were once AD, and the ARC depends on a steady stream of separating AD to fill its ranks. Many civilians are also former AD. Thus, any shift in manpower within these components could imbalance the flow of personnel to a point that becomes unsustainable. Civil engineering personnel expressed caution about the scale of firefighting manpower shifts we discussed: about 2,000 positions added to the AD force and about 2,000 positions cut from both the ARC and civilian workforces.[58] They cautioned that such a cut to the civilian workforce would mean the loss of significant expertise, expertise that has been leveraged to train AD personnel. Each career field has such dynamics that must be considered.

However, if a steady-state requirement is identified that demands a capability increase of that magnitude, the demand must be filled somehow. If it is not filled through a plus-up of AD forces, then the ARC may need to be accessed differently than is currently anticipated, or the AD D2D ratio will be more stressing than 1:2. In any case, expeditionary demands will often compete against such issues of corporate knowledge and expertise. Solving such problems will require insight into each career field's dynamics.

---

[56] Computed using FY12 Standard Composite Rates, from AFI 65-503.

[57] Conley et al., 2006.

[58] Email and phone discussions with Air Force Civil Engineering Center (AFCEC) personnel, March 2012.

Implementing Manpower Realignments

Undoubtedly, manpower realignments of the size and scope we consider here would take many years to implement. They would also not be without investment costs. Because we do not consider any net increase in AD end strength, only shifts among career fields, some implementation could be accomplished by simply shifting resources. Recruitment efforts could shift from one career field to another, as could training programs, bonuses and other incentives, etc. Also, some reductions could be accomplished organically as people retire or separate. But we realize that any significant change would entail much effort and certainly some investment cost. We estimated only the recurring savings, but if these concepts were to be implemented, the investment costs must also be estimated to assess the relative cost of any expeditionary capacity gained.

5. Conclusions and Recommendations

We now provide some concluding thoughts before recommending steps to implement these concepts.

There are Trade-Offs from Both an Enterprise and a Functional Perspective

In this report, we have compared the USAF's current system for shaping ACS manpower with our own approach. We argued that the USAF's current approach is more focused on functional areas and driven by home-station needs, while our approach is enterprise-focused and incorporates both expeditionary and home-station needs. There are advantages and disadvantages to both approaches.

A functionally driven approach appropriately retains focus on maintaining the health and viability of the career field and ultimately places control where most of the expertise currently lies. This approach takes into account accessions, training, career progression, and local base or major command (MAJCOM) issues. However, without policy or organizations that look across career fields and capabilities to provide a balanced portfolio, career fields may optimize capability within a function while suboptimizing the enterprise.

The manpower system's focus on home-station demands rather than expeditionary requirements is essentially a relic of the Cold War. This characteristic creates inherent imbalances between some ACS forces and operational forces when deployed, regardless of whether the capacity of these forces is compared to current or future-year plans. In contrast, a purely expeditionary-driven approach would likely impact the USAF's ability to organize, train, and equip its forces.

The important question is not which extreme is better, but what aspects of each approach are beneficial and ought to be retained. We suggest that the advantageous elements of our approach be implemented and, to the degree possible, integrated with the current system to reach a balanced approach to shaping the ACS force.

At root, we advocate changing the manpower planning system to incorporate the planning objectives we discussed above. However, we see two key obstacles to implementing the concepts we laid out and demonstrated in this report.

Obstacles to Implementing These Concepts

Authority and Organization

The USAF's current organization works against efforts to take an enterprise approach to balancing ACS capabilities. Today, there is no single organization or body that has directive authority over the entire ACS enterprise. While the functional capabilities of ACS contribute to coherent expeditionary capabilities, each functional community holds local expertise about its career field, and on the Air Staff, these ACS career fields fall under the management of many different directorates, including AF/A1, AF/A4/7, SAF/A6, SAF/AQ, and AF/SG. These stakeholders hold sway, as they set policy for their career fields.

Additionally, MAJCOM commanders are charged with the organize, train, and equip responsibilities that contribute to ACS capabilities. Mission support panels develop the programming inputs for many, but not all, ACS personnel, while mission panels are responsible for other ACS personnel. Finally, the ACS core function lead integrator (CFLI), as one of twelve CFLIs, develops programming inputs to integrate across ACS, but the ACS CFLI does not have directive authority over the entire enterprise and can, at best, only offer suggestions to those identified above. We understand that this structure is being changed to better align programming responsibilities with CFLIs, but how that process will work has not yet been determined.

If competing demands were to be arbitrated across all ACS functional capabilities, the Vice Chief of Staff of the Air Force would be the lowest-ranking position with the purview to do so. While the ACS CFLI has avenues through the Core Function Support Plan (CFSP), the planning force proposal, and more-detailed programming priorities to influence ACS resource decisions, the level of that influence will be determined by emerging changes to the planning and programming processes.

Doctrine, Policy, and Guidance

There is also a gap in strategic Air Force guidance, and as a result, functional communities have felt the need to fill this gap for their communities. Air Force Policy Directive (AFPD) 90-11 establishes the Air Force Strategic Planning System.[59] It identifies two key documents that could have a role in directing the USAF how to size and shape ACS. The first, the *Air Force Strategic Plan*, "sets goals and objectives for the Air Force in support of national and joint objectives and is the primary source document identifying priorities for the development and alignment of organizational strategic plans across the entire Air Force."[60]

Unfortunately, the USAF has not released a strategic plan in several years. To our knowledge, the most recent documents of this kind were released in 2006 (an official strategic plan)[61] and late 2007 (a white paper written by the Chief of Staff of the Air Force).[62] Much has changed since then. Further, neither document is a strategy in the sense of clearly linking ends,

---

[59] AFPD 90-11, *Strategic Planning System: Special Management*, Washington, D.C., March 26, 2009.

[60] AFPD 90-11, 2009.

[61] U.S. Air Force, *Air Force Strategic Plan 2006–2008*, Washington D.C., 2006.

[62] Michael T. Moseley, *The Nation's Guardians: America's 21st Century Air Force*, Washington, D.C., December 29, 2007.

ways, means, and risks.[63] These documents do not provide enough specificity or prioritization to guide decisionmaking in the planning and programming environments.

The recently released document, *Sustaining Readiness and Modernizing the Total Force*, does provide some specificity of a strategic planning nature, laying out objectives, a strategy to achieve those objectives, and fiscal choices to support that strategy.[64] This document has been the exception in recent years. We understand the USAF now plans to regularly develop an Air Force Strategic Master Plan to inform the CFLI support plans.

The second document, the *Annual Planning and Programming Guidance* (APPG), is "the primary document linking Air Force planning and programming." The APPG "outlines force structure and essential support requirements" and "provides programmers with prioritized guidance for capabilities and capacities that will be included in the program objective memorandum build and budgetary process."[65] This document has historically included some articulation of operational objectives in the form of planning scenarios.

However, each recent APPG has varied in its level of detail regarding specific planning scenarios and the degree of guidance given to programmers in how to trade off competing objectives when resources cannot support the full range of all stated planning objectives. One recent APPG had a single paragraph specifying planning scenarios to be used to shape the force, and these scenarios were not aligned with then-current OSD guidance documents on the topic. Recent APPGs have also not had a clear articulation of ACS objectives that could be used to size and shape the force. We are not advocating that all flexibility be removed from planning. We are simply advocating that enough detail be included in guidance to enable sufficient planning and programming to develop an agile and flexible force.

What is needed from USAF guidance is an articulation of operational objectives and priorities that is clear enough to be translated into quantifiable ACS objectives that can drive the size and shape of the future force. In this case, we refer to ACS objectives in the broadest sense, at the enterprise level, not necessarily the smaller scope of the current ACS service core function.

For several years, the ACS concepts of operations (CONOPS) laid out some ACS objectives to instruct functional areas in how much expeditionary capacity to posture (in the form of numbers of force modules for steady-state and surge activities). These CONOPS were not binding, however. While some functional communities did posture UTCs in accordance with the ACS CONOPS, this did not change the fundamental size or shape of ACS, only the packages of capability they presented. Ultimately, those CONOPS did not produce a force balanced against operational objectives.

---

[63] The 2007 white paper does mention ends, ways, means, and risks but spends about four pages describing these at an extremely high level.

[64] United States Air Force, *USAF Force Structure Changes: Sustaining Readiness and Modernizing the Total Force*, Washington D.C., March 2012.

[65] AFPD 90-11, 2009.

The Air Force Strategic Master Plan could articulate strategic and operational objectives to inform ACS requirements. Likewise, the ACS CFLI's CFSP could be a vehicle for establishing guidance around which to shape ACS forces, and its Planning Force Proposal (PFP) could drive those changes. However, because the CFLI currently lacks official directive authority over the ACS enterprise, any analysis the ACS CFLI presents will require broad consensus among the above stakeholders to be implemented. The emerging planning and programming structures may change that.

Potential Steps to Implement These Concepts

We make several recommendations for how to implement the concepts we outlined in this report.

Policy and Process

First, we recommend that the USAF provide clearer strategic planning guidance about future objectives to direct planners and programmers in sizing and shaping the force. While we described some of the perceived gaps in policy and guidance, we do not prescribe which is the appropriate vehicle to contain this guidance.

In this report, we have synthesized a range of defense planning scenarios (in a sense treating all of a given type as equally important) and illustrated several illustrative scenario sets as operational objectives. The recently released Defense Strategic Guidance changes those operational objectives and has already been translated into stated DoD and USAF force-structure reductions.[66] For USAF planners and programmers to tie resources to operational objectives, that OSD guidance must be translated into USAF objectives that can be quantified and analyzed.

Second, we recommend that those clearly stated USAF objectives be translated into measurable, enterprise-level ACS objectives. We used the type and number of bases as quantified ACS objectives and illustrated the outcome of that with alternative manpower mixes. There are certainly other important objectives, and they must be stated quantitatively to enable planners and programmers to shape the force.

Third, we recommend the ACS manpower system broaden its planning objectives to include expeditionary and home-station requirements. This would mean incorporating into the manpower system the capacity to assess expeditionary capabilities and balance them against home-station needs. This is no trivial task, as it requires making informed and sophisticated choices between efficiency and effectiveness. While the current total force enterprise (TFE)

---

[66] United States Air Force, 2012.

analysis seeks to do this to some extent, it is not yet clear how it will affect the manpower system.[67] This leads us to our final recommendation.

Fourth, we recommend that the USAF specify policies to inform manpower trade-offs when issues arise in balancing expeditionary vs. home-station needs or an enterprise vs. career-field perspective. For example, a career-field change that efficiently postures for expeditionary demands might compromise known needs for career progression or flow among the components. Such decisions must be informed from a high level because they involve the expeditionary capabilities the USAF will present, not just local career-management issues. These types of policy decisions must be made in order to implement any actions recommended by the ACS CFLI or the TFE analysis.

Some Analytic Skills and Expertise Are Needed

Beyond policy and process, several analytic skills or capabilities would be needed to meaningfully incorporate this kind of analysis into the USAF enterprise. The USAF already has these skills within its analytic communities, but to recreate analysis like that described in this report, this expertise would need to be better integrated and/or housed in one place to bring it to bear on these problems.

First, analysts need a basic understanding of defense plans and guidance. This is necessary to tie manpower to operational objectives in a meaningful way.

Second, analysts need a basic understanding of USAF UTCs and deployment concepts. If manpower planners and programmers are to design a force that can support USAF expeditionary operations, they need to understand how the myriad functional capabilities integrate to produce expeditionary capability.

Third, this task requires the ability to trade among and integrate competing objectives. As we state earlier, the USAF needs to provide policy to govern such strategizing (e.g., between functional and enterprise interests and expeditionary and home-station demands). It also requires the analytic capability to translate that policy into quantitative costs and benefits or constraints.

Finally, analysts need adequate modeling capability to integrate all these elements analytically and produce replicable results. For our analysis, we used Microsoft Excel spreadsheets and arithmetic calculations. AF/A9, for its support of the TFE analysis, uses an optimization approach. Whatever the approach, the organization must have a modeling capability that is stable and robust enough to support this type of analysis.

---

[67] The TFE analysis, initiated in 2010, is overseen by AF/A8XF, with the quantitative analysis performed by AF/A9RP. The purpose of the TFE analysis is to analyze total force manpower to illuminate trade-offs in cost, capability, and risk for senior leadership.

Further Research

While this report focuses on manpower, our methodology could be rather easily adapted to address the challenges of equipment also. Equipment does not have all the same complications manpower does, but in the same way that our methodology seeks to balance manpower capability across disparate areas, it could be used to create balance across equipment types and between equipment and manpower so that the USAF can maintain and present balanced capabilities.

Appendix

The purpose of this appendix is to explain in more detail the expeditionary ACS metrics we used in the body of this report. As shown in Figure 2.2 and explained in the accompanying text, we focused on surge and steady-state expeditionary operations. We looked to defense planning scenarios (their text and accompanying data sets) to determine the nature of the deployed locations in each of these contingency types. We found that two attributes drove most ACS manpower requirements: (1) the type and number of aircraft and (2) threat levels. We found that the number of aircraft, irrespective of type, correlated well with the total base population (which in turn drives many ACS manpower requirements). Thus, we use the number of aircraft, or squadron equivalents, at a base as a proxy for the base's size.

For surge demands, we used two long-range defense-planning scenarios from different theaters. Each scenario contains a beddown with the location and type and number of aircraft. From this, we derived a number of squadron equivalents of aircraft at each base. Figure A.1 shows how many squadrons those bases had. The x-axis shows the number of squadrons per base; the y-axis shows the frequency of occurrence of that base size in the two scenarios.

Figure A.1 Squadrons Per Base in Long-Range Surge Scenarios

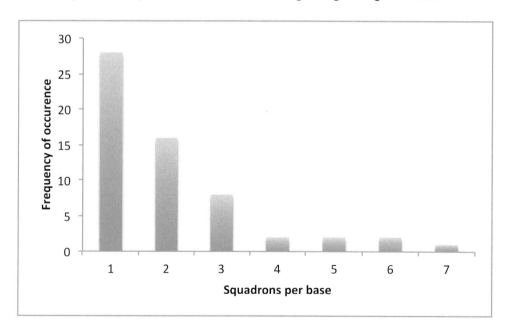

Together, these two surge scenarios had a total of 59 bases with USAF aircraft. About 75 percent of those bases had one or two squadrons, and the average number of squadrons was about 1.5 per base. The largest number of squadrons was seven, which occurred only once.

For steady-state scenarios, we used the defense planning scenarios from the Steady State Security Posture (SSSP) that was current in 2011. This set had dozens of scenarios and over 100 bases with USAF aircraft. Figure A.2 shows the frequency of different numbers of squadrons per base for these scenarios. This figure follows the same format as Figure A.1.

Figure A.2 Squadrons Per Base in Steady-State Scenarios

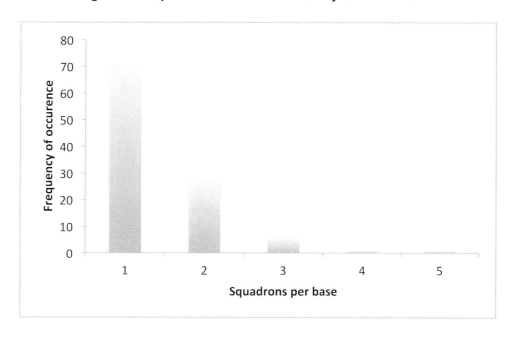

For this set of steady-state scenarios, two-thirds of the bases had only one squadron of USAF aircraft or less. Ninety-two bases had two squadrons or fewer. The average number of squadrons was about one per base. (Many of the bases shown in the first column of Figure A.2 had fewer than one squadron, e.g., twelve fighters).

We used the average squadrons-per-base numbers for surge and steady-state scenarios (1.5 and 1.0, respectively) to drive many of the ACS manpower requirements in the rest of our analysis.

These base metrics simplify what is, in reality, a range of base footprints. Taken in isolation, they do not capture the dynamics of some of the larger, less-common bases. This insight should caution the reader that these metrics are useful as rough tools for force planning, but not for more granular assessments.

Bibliography

AFI—*See* Air Force Instruction.

AFPD—*See* Air Force Policy Directive.

Air Force Instruction 10-401, *Air Force Operations Planning and Execution*, Washington, D.C., April 25, 2005.

Air Force Instruction 10-402, *Mobilization Planning*, Washington, D.C., May 1, 2012.

Air Force Instruction 38-201, *Management of Manpower Requirements and Authorizations*, Washington, D.C., September 26, 2011.

Air Force Instruction 65-503, *U.S. Air Force Cost and Planning Factors*, Washington, D.C., February 4, 1994. As of October 26, 2012:
http://www.e-publishing.af.mil/shared/media/epubs/AFI65-503.pdf

Air Force Personnel Center, UTA file, 2011.

Air Force Personnel Center, Data Retrieval Section, Active Enlisted End of Month Master Personnel Extract file, September 30, 2010.

Air Force Personnel Center, Data Retrieval Section, Authorized Manpower Master file, September 30, 2010.

Air Force Policy Directive 38-2, *Manpower*, Washington, D.C., March 2, 1995.

Air Force Policy Directive 90-11, *Strategic Planning System*, Washington, D.C., March 26, 2009. As of October 24, 2012:
http://www.af.mil/shared/media/epubs/AFPD90-11.pdf

Bennett, John T., "Thousands of US Troops Likely in Afghanistan Beyond 2014 Withdrawal Date," *U.S. News & World Report*, April 26, 2012. As of October 10, 2012:
http://www.usnews.com/news/blogs/dotmil/2012/04/26/thousands-of-us-troops-likely-in-afghanistan-beyond-2014-withdrawal-date

Congressional Budget Office, *Long-Term Implications of the 2012 Future Years Defense Program*, Washington, D.C., June 2011.

Conley, Raymond E., Albert A. Robbert, Joseph G. Bolten, Manuel Carrillo, and Hugh G. Massey, *Maintaining the Balance Between Manpower, Skill Levels, and PERSTEMPO*, Santa Monica, Calif.: RAND Corporation, MG-492-AF, 2006. As of October 24, 2012:
http://www.rand.org/pubs/monographs/MG492.html

Department of Defense Directive 1200.17, *Managing the Reserve Components as an Operational Force,* Washington, D.C., October 29, 2008. As of July 19, 2013: http://www.dtic.mil/whs/directives/corres/pdf/120017p.pdf

"Fact Sheet: The President's Framework for Shared Prosperity and Shared Fiscal Responsibility," The White House Office of the Press Secretary, Washington, D.C., April 11, 2011a. As of September 7, 2011: http://www.whitehouse.gov/the-press-office/2011/04/13/fact-sheet-presidents-framework-shared-prosperity-and-shared-fiscal-resp

Gates, Robert M., *Utilization of the Total Force: Memorandum for Secretaries of the Military Departments*, *Chairman of the Joint Chiefs of Staff, Under Secretaries of Defense,* January 19, 2007.

———, "Opening Summary—House Appropriations Committee–Defense (Budget Request)," Washington, D.C., March 2, 2011. As of November 1, 2012: http://www.defense.gov/Transcripts/Transcript.aspx?TranscriptID=4779

Greene, David, and Larry Abramson, "Panetta Makes an Unannounced Trip to Afghanistan," National Public Radio, June 7, 2012. As of October 10, 2012: http://www.npr.org/2012/06/07/154485144/leon-panetta-makes-an-unannounced-trip-to-afghanistan

"Iraqi Prime Minister Celebrates US Troop Withdrawal as New Dawn for Nation," Associated Press, December 31, 2011. As of December 31, 2011: http://www.washingtonpost.com/world/middle-east/iraqi-prime-minister-celebrates-us-troop-withdrawal-as-new-dawn-for-nation/2011/12/31/gIQANcwFSP_story.html

McKinley, Rodney, *Roll Call*, January 12–16, 2007. As of October 31, 2012: http://www.af.mil/shared/media/document/AFD-070112-002.pdf

Mills, Patrick, John Drew, Dan Romano, John Ausink, Mike McGee, Therese Bohusch, *Beyond Deploy-to-Dwell Ratios: New Metrics to Inform Home-Station Capability*, unpublished RAND research.

Mills, Patrick, David A. Shlapak, Ricardo Sanchez, and Robert S. Tripp, *Combat Support Beyond Iraq: Implications of the Future Security Environment for the USAF,* Santa Monica, Calif.: RAND Corporation, MG-1034-AF, 2011, not available to the general public.

Moseley, Michael T., *The Nation's Guardians: America's 21st Century Air Force*, Washington, D.C., December 29, 2007. As of December 10, 2013: http://www.google.com/url?sa=t&rct=j&q=&esrc=s&source=web&cd=2&ved=0CC4QFjAB&url=http%3A%2F%2Fwww.dtic.mil%2Fcgi-bin%2FGetTRDoc%3FAD%3DADA477488&ei=jimMUp6qKtKkkQfa7YGgCA&usg=AFQjCNFXdNf7Pubtp6lo1v7mj_l6MYlq-g&sig2=B-s5LN-qnM53G_LxMn8I9A&bvm=bv.56753253,d.eW0

Negrin, Matt, "The Troops in Iraq: Sent Home, as Promised," ABCNews.com, July 7, 2012. As of October 10, 2012:
http://abcnews.go.com/Politics/OTUS/troops-iraq-home-promised/story?id=16720328

"Obama Announces 34,000 Troops to Leave Afghanistan," *BBC News Online*, February 13, 2013. As of September 12, 2013:
http://www.bbc.co.uk/news/world-us-canada-21431423

Obama, Barack, "Remarks by the President on the Way Forward in Afghanistan," Washington, D.C., June 22, 2011. As of September 7, 2011:
http://www.whitehouse.gov/the-press-office/2011/06/22/remarks-president-way-forward-afghanistan

———, *Sustaining U.S. Leadership: Priorities for the 21st Century*, Washington, D.C., January 2012.

Public Law 112–25, Budget Control Act of 2011, Section 365, 125 Stat. 240, 2011).

Schwartz, Moshe, and Joyprada Swain, *Department of Defense Contractors in Afghanistan and Iraq: Background and Analysis*, Washington D.C.: Congressional Research Service, CRS Report for Congress, May 13, 2011.

Schwartz, Norton A., "Change to Air Expeditionary Force (AEF) Baseline," memorandum, Washington, D.C., September 2, 2010. As of September 7, 2011:
http://www.airforce-magazine.com/SiteCollectionDocuments/Reports/2010/September%202010/Day14/SchwartzPolicyLtr090210.pdf

———, "Air Force Strategic Choices and Budget Priorities Brief at the Pentagon," Washington, D.C., January 27, 2012. As of October 24, 2012:
http://www.defense.gov/Transcripts/Transcript.aspx?TranscriptID=4965

Snyder, Don, and Patrick Mills, *Supporting Air and Space Expeditionary Forces: A Methodology for Determining Air Force Deployment Requirements*, Santa Monica, Calif.: RAND Corporation, MG-176-AF, 2004. As of October 24, 2012:
http://www.rand.org/pubs/monographs/MG176.html

Snyder, Don, Patrick Mills, Manuel Carrillo, and Adam C. Resnick, *Supporting Air and Space Expeditionary Forces: Capabilities and Sustainability of Air and Space Expeditionary Forces*, Santa Monica, Calif.: RAND Corporation, MG-303-AF, 2006. As of October 24, 2012:
http://www.rand.org/pubs/monographs/MG303.html

United States Air Force, *Air Force Strategic Plan 2006–2008*, Washington D.C., 2006. As of December 10, 2013:
http://www.au.af.mil/au/awc/awcgate/af/af_strat_plan_06-08.pdf

———, *Annual Planning and Programming Guidance*, Washington, D.C., 2011a, not available to the general public.

_____, *USAF Strategic Planning 2010–2030 Strategic Environmental Assessment*, March 11, 2011b.

_____, *USAF Force Structure Changes: Sustaining Readiness and Modernizing the Total Force*, Washington D.C., March 2012.

United States Department of Defense, *Quadrennial Defense Review Report*, Washington, D.C., September 30, 2001. As of October 28, 2012:
http://www.defense.gov/pubs/qdr2001.pdf

_____, *Defense Planning Scenario: Steady-State Security Postures/Integrated Security Postures (SSSP/ISP)*, Final, Washington, D.C., April 18, 2008a, not available to the general public.

_____, *Managing the Reserve Components as an Operational Force*, Department of Defense Directive 1200.17, Washington, D.C., October 29, 2008b.

_____, *Quadrennial Defense Review Report*, Washington, D.C., February 2010a.

_____, *Department of Defense (DoD) Efficiency Initiatives*, Washington, D.C., August 16, 2010b.